(*continued on back*)

Photochemical Vapor Deposition

CHEMICAL ANALYSIS

A SERIES OF MONOGRAPHS ON
ANALYTICAL CHEMISTRY AND ITS APPLICATIONS

Editor
J. D. WINEFORDNER
Editor Emeritus: **I. M. KOLTHOFF**

VOLUME 122

A WILEY-INTERSCIENCE PUBLICATION

JOHN WILEY & SONS, INC.

New York / Chichester / Brisbane / Toronto / Singapore

Photochemical Vapor Deposition

J. G. EDEN

Everitt Laboratory
Department of Electrical and Computer Engineering
University of Illinois
Urbana, Illinois

A WILEY-INTERSCIENCE PUBLICATION

JOHN WILEY & SONS, INC.

New York / Chichester / Brisbane / Toronto / Singapore

Seplae
chem

Copyright © 1992 by John Wiley & Sons, Inc.

Library of Congress Cataloging in Publication Data:

Eden, J. G.
    Photochemical vapor deposition / J. G. Eden.
      p. cm.—(Chemical analysis, ISSN 0069-2883 ; v. 122)
    "A Wiley-Interscience publication."
    Includes bibliographical references and index.
    ISBN 0-471-55083-3 (alk. paper)
    1. Vapor-plating. 2. Thin films. 3. Photochemistry. I. Title.
  II. Series.
  TS695,E33 1992   92-24564
  671.7'35—dc20

10 9 8 7 6 5 4 3 2 1

Dedicated to my parents,
Bob and Jodie Eden,
and to my grandmothers,
Bertie West and Forest Eden

PREFACE

Periodically updating the progress in a rapidly moving technical area often has a positive impact on its further development by providing a focus for research efforts and in presenting a unified body of work to those who are unfamiliar with the field but yet potentially able to benefit from its results. This book is offered in the hope that it will provide a useful overview of photochemical vapor deposition (photo-CVD), its characteristics and potential by emphasizing the properties of films that have been grown by this versatile new technique and through a description of the conditions under which they were grown. The subject is approached from the premise that while photo-CVD is not the solution to all low-temperature deposition applications, it offers an added dimension of flexibility to and control over the growth process and has clearly proven to be valuable for those materials and growth steps in device fabrication that are sensitive to the processing temperature.

This book has been prepared with several audiences in mind: the student approaching photodeposition for the first time, the thin-film engineer wishing to evaluate photo-CVD for a particular application and possibly adapt it into an existing process, and the experienced researcher desiring a review of the work completed to date. For the newcomer to the field, introductory material describing the basic principles underlying photo-CVD is given in Chapters 1 and 2. Chapter 3 addresses several practical aspects of designing photo-CVD systems (and choosing an optical source, in particular), and Chapters 4–6 discuss the current status of metal, semiconductor, and dielectric films, respectively, that have been grown by photo-CVD. Several other materials (including polymers) and the application of photo-CVD films to the fabrication of multilayer, short wavelength mirrors are described in Chapter 7, and Chapter 8 concludes by venturing a forecast concerning the future of photodeposition. Whenever applicable, the properties of electronic devices incorporating photo-CVD films are presented, and tables throughout the text describe in more detail the deposition parameters and electrical and structural properties of the films of specific materials. For further detail concerning various facets of photo-CVD, several excellent review articles and books are referenced in the text for the interested reader. Although intended

vii

to be a convenient reference source, this book is not exhaustive and the discussion emphasizes the photodeposited films of specific materials such as Cr, Si, SiO_2, and the Group II–VI compound semiconductors that have been studied most extensively and that stand out as having especially benefited from the low-temperature capabilities of photo-CVD.

It is also perhaps inevitable that aspects of a book such as this become dated rapidly because of ongoing developments in the deposition of new materials, synthesis of more effective precursors, and the engineering of optical sources and hybrid surface treatments. This is as it should be, but the foundational principles undergirding photodeposition will remain unchanged. Pointing out threads in past work and proposing areas most in need of further effort will hopefully stimulate interest in this exciting area. The reader must judge whether this attempt has been successful.

The author owes a considerable debt to a number of individuals whose support and encouragement were instrumental in bringing this book to fruition. Specifically, many discussions with and suggestions from F. A. Houle, J. J. Coleman, J. T. Verdeyen, T. Donohue, J. E. Greene, S. D. Allen, I. P. Herman, S. J. C. Irvine and R. M. Osgood, Jr., are appreciated. Also, the support of H. R. Schlossberg of the Air Force Office of Scientific Research and the National Science Foundation Division of Materials Research through the University of Illinois Materials Research Laboratory (under grant NSF DMR 89-20538) throughout the preparation of the manuscript is gratefully acknowledged. Kelly Voyles and Jean Sexton exhibited tremendous typing and proofreading skills in readying the text for publication, and for that, as well as their continual enthusiasm, I am most appreciative. My deepest thanks are reserved for my wife Carolyn for her extraordinary patience and Rob, Laura, and Kate for their good-natured toleration of a busy and often absentminded dad during this time.

J. G. EDEN

Urbana, Illinois
June 1992

CONTENTS

Photochemical Vapor
Deposition

CHAPTER

1

INTRODUCTION

*The Light (φῶς) shines in darkness and the darkness
did not overcome it.*

JOHN 1 : 5

Photochemistry is the science concerned with the pursuit of various approaches to modifying or initiating chemical reactions with light. As a natural
consequence of the insight into the fundamental structure of matter it has
provided, photochemistry has spawned applications as diverse as the synthesis
of vitamin D and the separation of atomic isotopes with lasers. Although the
deposition of films by gas phase photochemical reactions was mentioned in
the scientific literature more than 50 years ago (1-4), only since the late 1970s
has the growth of high-quality elemental and compound films by photochemical processes been developed vigorously. The expansion of the variety of
materials that can be deposited by photochemical processes is truly remarkable: currently, photo-assisted chemical vapor deposition (photo-CVD) en-
compasses 25 elements of the periodic chart, in addition to at least 20 dielectric
and semiconductor compound films.

Conventional chemical vapor deposition (CVD) entails decomposing one
molecule or a combination of polyatomic vapors entrained in a gas flow
stream by passing the mixture over a heated substrate. Since the decomposition of the molecular reactants (precursors) is thermally activated, substrate
temperature is the single most important system parameter. Despite the fact
that CVD has proven to be successful in growing a vast array of material films
(5) and undergirds the microelectronics industry in the deposition of dielectric
and semiconductor films, further flexibility in the deposition process (including a reduction in the necessary film growth temperature) can be realized by
the introduction of light into the thin film reactor.

Interest in photo-CVD partially stems from the ability of optical radiation
to induce specific chemical reactions in the gas phase or at a surface. In the
visible, ultraviolet (UV), and vacuum ultraviolet (VUV) regions of the elec-
tromagnetic spectrum, photons have energies comparable to or exceeding
those of most chemical bonds. It is therefore not surprising that the interaction

of optical radiation with a polyatomic molecule is capable of selectively rupturing bonds within the molecule and yielding specific products in the form of a thin film. The selective optical production of atoms, molecular radicals, or even excited species in the vicinity of a surface and the ability to do so independently of the substrate temperature effectively decouple temperature from the number density of the species of interest. Said another way, the introduction of photons into a thin film deposition reactor allows one to drive the chemical environment far from equilibrium by selectively producing species that are not normally present in significant concentrations in conventional CVD, molecular beam epitaxy (MBE), or metal organic chemical vapor deposition (MOCVD) reactors. This flexibility inherent with the photodeposition of films often permits operation at lower temperatures where impurity redistribution and thermally induced mechanical stress are minimized (6–8). Other attractive aspects of photo-CVD include the capability for selective area deposition and, in general, the absence of ion damage to the film that often accompanies plasma-assisted deposition techniques. In short, the spatial resolution, chemical specificity, and reduced temperature capability of photodeposition are among the factors driving the current efforts in this field.

Despite the fact that the development of photo-CVD has been in progress for only a decade, dramatic progress in film quality and in the variety of techniques available for deposition has been made. Over the last 7 years in particular, the structural, chemical, and electronic properties of photo-CVD-deposited dielectric films, such as defect generation, step coverage, and uniformity, have improved steadily, prompting the introduction of commercial reactors, the incorporation of photo-CVD SiO_2 layers into VLSI (very-large-scale integration) processing (9, 10), and the utilization of photodeposited films in the fabrication of III–V and II–VI semiconductor devices (6, 11, 12).

It must be stressed that photodeposition techniques are still in the early stages of development. The feasibility of growing a wide range of materials has been demonstrated, but the painstaking effort of engineering the process for each film to ensure the reproducible production of high-quality material lies ahead. Nevertheless, the flexibility in controlling surface and gas phase reactions that light affords is too versatile to not be exploited, and photo-CVD will undoubtedly play an increasing role in film deposition technology in the future.

It is the goal of this monograph to (1) provide a broad overview of the photodeposition methods and materials that have been reported through the fall of 1991, and (2) illustrate the potential of emerging photo-CVD techniques to low-temperature processing applications. Emphasis will be placed on the application of visible or shorter wavelength radiation from both lasers and lamps to the photodeposition of films of interest to microelectronics. Photo-CVD can be viewed as a specialized branch of photochemistry that is preoccupied with generating a specific class of products, namely, low-volatility materials comprising metal, dielectric, and semiconductor films for commercial applications. Of course, the desired products dictate the precursors that one is likely to explore. For this reason, much of the discussion of Chapter 2 concerning the optical processes most commonly applied to the photodeposition of films will be familiar to the experienced photochemist, but the treatment will be skewed toward the key photoprocesses occurring in molecules that are both volatile and include one or more of the constituent atoms in a desired film.

Because of their similarity to conventional CVD techniques, those photodeposition processes that are primarily thermal in nature will not be treated here. Such *photothermal* processes rely on the spatial properties of laser radiation to locally heat a substrate, and the reaction zone is laterally confined by the substrate's thermal conductivity and the nonlinear dependencies of reaction rates on temperature. Deposition processes involving the multiphoton absorption of infrared radiation to thermally dissociate polyatomic molecules will also not be discussed. Only rarely, however, is one able to attribute a particular photodeposition process to either photochemical or photothermal processes exclusively, and the impact of the two classes of reactions on the deposition of specific films will be noted whenever possible.

Photo-CVD is only one member of an ever-growing class of processes, including those fundamentally photochemical *or* thermal in nature, that are now accessible with optical radiation. Laser-assisted doping, for example, is only briefly discussed here in conjunction with specific epitaxial semiconductor films, and the rapidly expanding and fertile field of laser-induced etching of films will not be reviewed specifically although photo-assisted etching frequently draws upon those photochemical techniques that underlie photo-CVD. For the interested reader, excellent reviews of laser-assisted etching and photothermal processing techniques can be found in Vossen and Kern (5). Also, several excellent reviews of photo-assisted deposition of semiconductor, metal, and insulator films have appeared in the literature. Early work in this field was summarized by Osgood (13), and over the past 7 or 8 years a half-dozen journal articles (14–19) and several chapters or entire books (20–26) devoted to reviewing the fundamental chemical and physical aspects of laser microchemical processing have been published.

Our discussion of photodeposition begins with a review of the fundamental chemical mechanisms that are crucial to determining the properties of the resulting films. Although the three media available for interaction with the optical beam (gas or vapor, adlayer, or the substrate bulk) will be discussed, the various photochemical approaches (processes) available for deposition based on interactions with the gas phase or the adlayer will be stressed. Chapter 3 reviews several aspects of reactor design and discusses the properties of the optical sources (coherent and incoherent) that are now available in the visible, UV, and VUV regions.

Chapters 4–6 summarize the current status of the photodeposition of metal, semiconductor, and dielectric thin films. Despite the variety of films that have been grown successfully, the development of photo-CVD as a reliable and well-understood tool for the device engineer or material scientist is in its early stages. Consequently, a thorough discussion of the structural and chemical properties of deposited films will be reserved for those materials such as Si, SiO_2, and Cr that have received the most attention and for which a fair comparison of the photo-CVD film properties with those deposited by other, further developed techniques can be made. Chapter 7 provides a brief review of other materials, including polymers and superconducting films, that have been deposited photochemically and for which the low-temperature capability of photo-CVD is a tremendous asset. Chapter 8 discusses several benefits (such as surface cleaning) associated with photochemical processing, and the book concludes with an assessment of the potential of photo-CVD in the near term (5 years).

CHAPTER

2

FUNDAMENTAL ASPECTS OF PHOTOCHEMICAL VAPOR DEPOSITION

*... may not bodies receive much of their activity from the
particles of light which enter their composition?*

ISAAC NEWTON (*Optiks*, 1730)

2.1. PHOTOTHERMAL AND PHOTOCHEMICAL PROCESSES: GENERAL CLASSIFICATION OF PHOTO-ASSISTED DEPOSITION PROCESSES

Photo-assisted film deposition methods and laser-assisted deposition techniques, in particular, may be broadly subdivided into one of several categories, depending upon which of the three media (or combination thereof) present in a reactor absorbs strongly at the laser wavelength. As illustrated in Figure 1, (27), the substrate is normally immersed in a gas or mixture of gases that gives rise to the formation of an adlayer at the surface of the substrate (to be discussed in Section 2.3 of this chapter). At the core of all photo-assisted deposition processes is the gas, vapor, or mixture of gases, from which we wish to extract the element(s) of interest. Known as the precursor or "parent," the gas is generally molecular and is chosen for its combination of thermochemical and spectral properties. All photodeposition methods can be qualitatively distinguished from one another according to the manner in which the precursor is decomposed to yield the desired film.

Depending upon the wavelength and intensity of the optical source, either the gas, adlayer, or substrate will be the predominant absorber which has a critical bearing on the physical characteristics and growth rate of the resulting film. The first question to be considered when attempting to categorize photo-assisted deposition processes is, how is the absorbed energy disposed of? If the optical source energy is immediately converted primarily into heat, then the film growth process is said to be photothermal. Also known as laser chemical vapor deposition (LCVD), this approach is based on the laser providing a spatially localized source of heat at the substrate that decomposes (by pyrolysis, from the Greek word for burning, *purōsis*) the precursor molecules

5

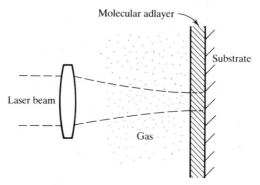

Figure 1. Diagram illustrating the three media—gas, adlayer, and substrate—that are generally present in a photodeposition reactor [after Ehrlich et al. (*27*), reprinted by permission]. The optical and chemical properties of these three play a key role in determining the nature of the reactions that culminate in film growth and in the metallurgical and electrical properties of the deposit. Although the laser is shown irradiating the substrate, films are also often grown with the optical beam directed parallel to the surface (see Figure 3b).

once the surface temperature exceeds a critical value that varies from molecule to molecule. In this case, the gas in the reactor is often transparent at the laser wavelength and, except for the heat source, LCVD is otherwise similar to conventional CVD. LCVD places few constraints on the wavelength of the laser, and the lateral extent of film growth in the plane of the substrate's surface is determined not only by the cross-sectional geometry of the laser beam itself but also on the thermal conductivity of the substrate and nonlinear variation of the pyrolytic reaction rate constant with surface temperature. A second variation of LCVD depends on the gas of interest absorbing the laser radiation in such a way that the precursor molecules are not dissociated optically but are heated and eventually fragmented by thermal processes. Whether the substrate or gas (or both) are heated by the optical source, the results are similar and such processes are categorized as photothermal.

Photochemical vapor deposition, on the other hand, owes its existence to the fact that UV and visible photon energies (~ 2–$6\,eV$) are often sufficiently large that chemical bonds in polyatomic molecules can be ruptured by the absorption of one or perhaps two photons. Such processes dictate the maximum allowable wavelength (or minimum photon energy) of the optical source and are characterized by the disposal of a significant fraction of the energy absorbed from the optical source into the breaking of chemical bonds and, hence, fragmentation of the precursor molecule.

A wide range of photo-assisted deposition techniques have been demonstrated and, bearing the above principles in mind, most are readily classified

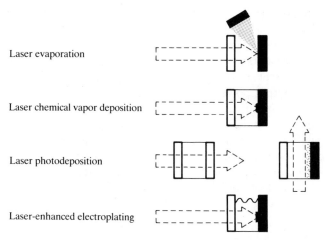

Figure 2. Schematic representations of the various methods available for the laser-assisted deposition of films [after Houle and Allen (28), reprinted by permission]. The darkened rectangles represent substrates that absorb strongly at the laser wavelength, whereas the open rectangles are transparent materials. Note that laser-enhanced electroplating occurs in the presence of a liquid. This book will be directed toward the third area, laser (or lamp) photodeposition of thin films.

as shown schematically in Figure 2 (28). In this diagram, substrates that absorb strongly at the laser wavelength are denoted by solid rectangles whereas optically transparent materials are represented by open rectangles. Laser evaporation consists of rapidly heating a target with a pulsed laser, which results in the ablation of surface molecules and their subsequent deposition onto a nearby substrate. In this case, the gas (if present) is nonreactive and frequently the process is carried out in vacuum. A reactive ambient gas is essential for laser CVD, however, and the processes that culminate in film growth are heterogeneous in that they occur at the substrate surface. Laser-enhanced electro-plating requires the presence of a liquid at the substrate and will not be discussed further, although other liquid-based deposition processes will be briefly touched on in Chapter 4 (Section 4.6). The remainder of this chapter will concentrate on film growth processes based predominantly on photochemical interactions with molecules in the gas phase or in the adlayer.

The reactor configurations generally used in photochemical vapor deposition can, as depicted in Figure 2, involve an absorbing or optically transparent substrate, and it is not necessary for the optical beam to actually irradiate the surface. Both the so-called perpendicular (or normal incidence) and parallel geometries common to photochemical vapor deposition are illustrated in more detail in Figure 3. As its name implies, perpendicular geometry involves irradiating the substrate directly, and the critical photochemical mechanisms in this configuration generally occur in the adlayer. While Figure 3 illustrates a laser as the optical source, a large fraction of photodeposition processes yield excellent results with UV lamps and, because of their lower instantaneous output powers, one is able to directly irradiate the substrate as in Figure 3a without concern for significantly heating the surface by the optical source. In parallel geometry, the optical beam does not impinge on the substrate but rather produces atomic or molecular species that migrate to the surface and react with the surface and adlayer, resulting in the desired film. In both cases (perpendicular or parallel geometry), the "active" medium of Figure 3 is, in fact, initially inactive in the sense that no film is grown

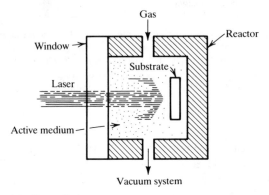

(a)   Perpendicular

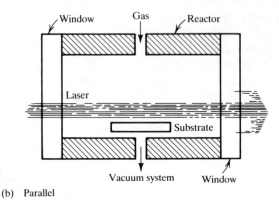

(b)   Parallel

Figure 3. (a) Perpendicular (or normal incidence) and (b) parallel geometry configurations typically used in laser photodeposition. Because of their lower intensities, UV lamps can irradiate the substrate (configuration a) without inducing a significant rise in surface temperature.

until the optical radiation photochemically alters the molecules (in the adlayer or gas phase) into a more chemically reactive form. One of the attractive features of photochemical deposition, then, is the control that it offers— spatial and temporal—since, in the absence of the external radiation provided by the lamp or laser, the film growth rate is negligible or zero.

In summary, the single most important factor in determining the nature of any photo-assisted deposition process is the absorption spectra of the three media (gas, adlayer, and substrate) and the reaction channels that predominate at a specified wavelength. Occasionally, it is difficult to resolve the contributions of photochemical and thermal mechanisms to the overall growth process; in fact, intentionally combining the two may be advantageous in some situations. As an example, the photochemical production of a molecular radical from the precursor can be followed by the photothermal decomposition of the *radical* at the surface: (1) as a means of producing a higher quality film; (2) in an effort to lower the required substrate temperature; or (3) as an approach to using molecular precursors of lower toxicity. A deciding factor in integrating the two growth mechanisms for a specific application may be that photothermal (pyrolytic) deposition rates frequently exceed those for its photochemical counterpart.

During the course of growing a film in perpendicular geometry (Figure 3a), the dominant growth mechanism can switch from photochemical to thermal. If the substrate does not absorb appreciably at the optical source wavelength, growth is initiated by photochemically produced species interacting with the surface. As time progresses, however, the growing film generally does absorb and so thermal processes can gain the upper hand after the film reaches a few tens of nanometers in thickness.

2.2. OVERVIEW OF PHOTOCHEMICAL PROCESSES

Photochemistry in the gas and liquid phases is a well-established science, and considerable effort has been devoted over the last half century, in particular, to determining the products formed when small molecules absorb a visible, UV, or VUV photon (photon energy, $\hbar\omega \simeq 2$–$10\,eV$; bond breaking with radiation of longer wavelengths [infrared (IR)] typically requires the absorption of more than one photon). Most of this information has been obtained from gas phase experiments and considerably less is known of the interactions of optical radiation with molecular adlayers. This chapter will review those photochemical processes that are most often at the heart of photodeposition techniques. For further detail concerning gas phase and adlayer fundamentals, the reader is encouraged to consult Refs. *29–31* and *19–24*, respectively.

The absorption of optical radiation by a gas of number density N (expressed in cm^{-3}) over a path length L is, at low intensities, governed by the relation (Beer–Lambert law):

$$I_t = I_i \exp[-N\sigma L] \tag{1}$$

where I_i and I_t are the optical intensities ($W \cdot cm^{-2}$) incident upon and transmitted by the gaseous medium, respectively, and σ is the absorption cross section (expressed in cm^2) for the gas in question. Often in the literature, one will find the absorption coefficient given as k and expressed in units of $atm^{-1} \cdot cm^{-1}$ (or $Torr^{-1} \cdot cm^{-1}$, where 1 Torr is the equivalent of a number density of $3.21 \cdot 10^{16} \, cm^{-3}$ at 300 K). At 0°C (273 K), the relationship between σ and k is (29)

$$\sigma = 3.72 \cdot 10^{-20} k \tag{2}$$

Also, the molar absorption (or extinction) coefficient ε is expressed in units of $liters \cdot mol^{-1} \cdot cm^{-1}$ and at 0°C is defined as (29)

$$\varepsilon = 9.73 k \tag{3}$$

or, in more fundamental terms,

$$\varepsilon = -\frac{1}{CL} \log_{10}\left(\frac{I_t}{I_i}\right) = \frac{A}{CL} \tag{4}$$

where C is the concentration of the molecule (in mol/liter), and A is known as the absorbance. Although not explicitly indicated above, σ, k, and ε vary with wavelength λ and are therefore often written as $\sigma(\lambda)$, $k(\lambda)$, and $\varepsilon(\lambda)$. Equations (1)–(4) are valid only when the precursor molecule absorbs a single photon from the optical source beam; multiphoton absorption rates, which vary nonlinearly with the source intensity I and are generally significant only for $I \gtrsim 10^5$–$10^6 \, W \cdot cm^{-2}$, will be discussed later.

The fraction of the laser or lamp beam energy absorbed by the gas hinges upon the value of σ at the source wavelength. For example, given a precursor partial pressure of 1 Torr and an absorption cross section of $10^{-18} \, cm^2$, 3% of the optical source energy will be absorbed over a 1-cm path length. If $\sigma = 10^{-17} \, cm^2$ at the source wavelength, the absorbed power rises to 27%. Since the absorption cross section is a function of the source wavelength, measuring the visible, UV, and VUV absorption *spectra* for volatile precursors containing metal or semiconductor atoms is the first consideration in assessing the utility of a polyatomic molecule for photodeposition. Table 1

Table 1. Photoabsorption Cross Sections at Several Wavelengths (λ, nm) in the UV for Polyatomic Molecules Useful as Precursors for Photo-CVD[a]

Precursor Molecule	Element of Interest	$\sigma(\lambda)$, cm^2			Reference(s)
		$\lambda = 193$ nm	248 nm	308 nm	
Al(CH$_3$)$_3$	Al	$(2.0 \pm 0.9) \cdot 10^{-17}$	$1.1 \cdot 10^{-20} - 8 \cdot 10^{-19}$		27,73–76
Al$_2$(CH$_3$)$_6$	Al	$(2.0 \pm 0.3) \cdot 10^{-17}$	$1 \cdot 10^{-19}$		76,77
Al(C$_2$H$_5$)$_3$	Al	$(4.7 \pm 2.0) \cdot 10^{-18}$			78–80
Al(i-C$_4$H$_9$)$_3$	Al	$1.7 \cdot 10^{-17}$			80
AlH(CH$_3$)$_2$	Al		$7 \cdot 10^{-20}$		81
AsH$_3$	As	$1.8 \cdot 10^{-17}$			82
As(CH$_3$)$_3$	As	$(4.8 \pm 0.5) \cdot 10^{-17}$			76,83
As(C$_2$H$_5$)$_3$	As	$1.8 \cdot 10^{-17}$			83
BCl$_3$	B	$5.4 \cdot 10^{-20}$			84
B$_2$H$_6$	B	$4.3 \cdot 10^{-20} - 2.2 \cdot 10^{-19}$			82,85
B(CH$_3$)$_3$	B	$5 \cdot 10^{-19}$ (191 nm)			86
B(C$_2$H$_5$)$_3$	B	$4.4 \cdot 10^{-19}$			87
CH$_4$	C	$<10^{-21}$			29
C$_2$H$_4$	C	$1.5 \cdot 10^{-20} - 1.9 \cdot 10^{-19}$			88–90
Cd(CH$_3$)$_2$	Cd	$(7.0 \pm 2.5) \cdot 10^{-18}$	$(4.0 \pm 1.0) \cdot 10^{-18}$		27,34,38,91,92
Cr(CO)$_6$	Cr	$(1.9 \pm 0.7) \cdot 10^{-17}$	$(4.8 \pm 1.0) \cdot 10^{-17}$	$(5.3 \pm 0.2) \cdot 10^{-18}$	58,93–95
CrO$_2$Cl$_2$	Cr		$3.0 \cdot 10^{-18}$	$5.1 \cdot 10^{-18}$	96
Fe(CO)$_5$	Fe	$5.7 \cdot 10^{-17}$	$2.7 \cdot 10^{-17}$		97–99
Ga(CH$_3$)$_3$	Ga	$(2.2 \pm 0.6) \cdot 10^{-17}$	$(2.2 \pm 1.2) \cdot 10^{-18}$	$(1.8 \pm 0.5) \cdot 10^{-18}$	73,75,78,79, 83,86,100
Ga(C$_2$H$_5$)$_3$	Ga	$(6.8 \pm 1.8) \cdot 10^{-18}$	$(5.6 \pm 2.0) \cdot 10^{-19}$		75,78,79,83
GeH$_4$	Ge	$(2.5 \pm 0.5) \cdot 10^{-20}$	$6 \cdot 10^{-23}$		33,101
Hg(CH$_3$)$_2$	Hg	$(2.6 \pm 0.1) \cdot 10^{-17}$	$7.1 \cdot 10^{-20} - 3.2 \cdot 10^{-19}$		91,92,102

11

Table 1 (*Continued*)

Precursor Molecule	Element of Interest	$\sigma(\lambda)$, cm²			Reference(s)
		$\lambda = 193$ nm	248 nm	308 nm	
$In(CH_3)_3$	In	$(1.2 \pm 0.2) \cdot 10^{-17}$	$(2.7 \pm 1.5) \cdot 10^{-18}$		35, 73, 86, 103
$In(CH_3)_2C_2H_5$	In	$1.2 \cdot 10^{-17}$	$(3 \pm 1) \cdot 10^{-18}$	$3 \cdot 10^{-19}$	76
InI	In	$4 \cdot 10^{-16}$			71
$(CH_3)_3InP(CH_3)_3$ (adduct)	InP		$7.8 \cdot 10^{-18} (240 \text{ nm})$		104
$Ir(acac)_3$	Ir		$5.9 \cdot 10^{-18}$		39
MoF_6	Mo	$4.9 \cdot 10^{-18}$	$2.0 \cdot 10^{-19}$	$< 10^{-21}$	105, 106
$Mo(CO)_6$	Mo	$(5.5 \pm 0.4) \cdot 10^{-17}$	$(5.1 \pm 0.7) \cdot 10^{-17}$	$(1.2 \pm 0.1) \cdot 10^{-17}$	94, 107
NH_3	N	$(1.1 \pm 0.5) \cdot 10^{-17}$			46–48, 50
$Ni(CO)_4$	Ni			$2.4 \cdot 10^{-18}$	108
N_2O	O	$(8.6 \pm 0.7) \cdot 10^{-20}$			109, 110
NO_2	O	$(4.5 \pm 1.9) \cdot 10^{-19}$	$(2.4 \pm 1.0) \cdot 10^{-20}$	$1.6 \cdot 10^{-19}$	111, 112
O_2	O	$1.4 \cdot 10^{-21}$			90
PH_3	P	$(1.3 \pm 0.1) \cdot 10^{-17}$			82, 113, 114
$P(CH_3)_3$	P	$3.4 \cdot 10^{-17}$			103
$PH_2(CH_3)_3C$	P		$1 \cdot 10^{-19}$		76
$P(C_2H_5)_3$	P	$(2.1 \pm 0.1) \cdot 10^{-17}$			103
$Pb(CH_3)_4$	Pb	$8.5 \cdot 10^{-18}$	$3.7 \cdot 10^{-19} - 9.6 \cdot 10^{-18}$ (257 nm)		115, 116
$\eta^5\text{-}C_5H_5Pt(CH_3)_3$	Pt			$6.1 \cdot 10^{-18}$	117
$Pt(acac)_2$	Pt		$2.4 \cdot 10^{-18}$		39
$Pt(CFacac)$	Pt		$7.2 \cdot 10^{-18}$ (260 nm)		118
ReF_6	Re	$7.3 \cdot 10^{-18}$			119
$S(CH_3)_2$	S	$6 \cdot 10^{-18}$	$< 10^{-19}$		38

12

$S(C_2H_5)_2$	S	$1.1 \cdot 10^{-17}$		$< 5 \cdot 10^{-19}$		38
$Sb(CH_3)_3$	Sb	$2.8 \cdot 10^{-17}$		$1.7 \cdot 10^{-18}$		35
$Se(CH_3)_2$	Se	$1.1 \cdot 10^{-17}$		$1.5 \cdot 10^{-18}$	$2 \cdot 10^{-19}$	38
$Se(C_2H_5)_2$	Se	$1.9 \cdot 10^{-17}$		$2 \cdot 10^{-18}$	$3 \cdot 10^{-19}$	38
SiH_4	Si	$1.1 \cdot 10^{-21}$				82
$Si(CH_3)_4$	Si	$< 10^{-21}$				120
Si_2H_6	Si	$4 \cdot 10^{-18}$ (190 nm)				32
$Si_2(CH_3)_6$	Si	$4.6 \cdot 10^{-17}$				120
Si_3H_8	Si	$3.4 \cdot 10^{-17}$ (190 nm)				32
$SiH_3(C_6H_5)$	Si	$3.0 \cdot 10^{-16}-3.8 \cdot 10^{-17}$ (190 nm)				121,122
$SnCl_4$	Sn	$3.8 \cdot 10^{-17}$ (201 nm)	$8.3 \cdot 10^{-18}$			123,124
$Sn(CH_3)_4$	Sn	$4.0 \cdot 10^{-17}$ (186 nm)	$1.9 \cdot 10^{-22}$ (254 nm)			118,123
$TaCl_5$	Ta		$1.9 \cdot 10^{-18}$ (254 nm)			125
$Ta(OCH_3)_5$	Ta		$4.1 \cdot 10^{-19}$ (254 nm)			126
$Te(CH_3)_2$	Te	$3.8 \cdot 10^{-18}-2.8 \cdot 10^{-17}$	$(2.9 \pm 1.0) \cdot 10^{-17}$		$5 \cdot 10^{-19}$	38,127–129
$Te_2(CH_3)_2$	Te		$2.7 \cdot 10^{-18}$			129
$Te(C_2H_5)_2$	Te	$(3.2 \pm 1.6) \cdot 10^{-17}$	$(3.8 \pm 2.0) \cdot 10^{-17}$		$5 \cdot 10^{-19}$	38,91,128,129
$TiCl_4$	Ti	$3 \cdot 10^{-17}$	$1.1 \cdot 10^{-17}$		$2.3 \cdot 10^{-18}$	130,131
$TiBr_4$	Ti		$2.0 \cdot 10^{-17}$		$1.2 \cdot 10^{-18}$	130
TlI	Tl	$4.5 \cdot 10^{-16}$	$1.4 \cdot 10^{-17}$			66,132
VCl_4	V		$1.0 \cdot 10^{-17}$		$1.1 \cdot 10^{-17}$	133
$VOCl_3$	V		$1.8 \cdot 10^{-17}$		$5.0 \cdot 10^{-18}$	134
WF_6	W	$3.5 \cdot 10^{-19}$				105
$W(CO)_6$	W	$1.2 \cdot 10^{-17}$	$4.5 \cdot 10^{-18}$		$2.4 \cdot 10^{-18}$	94
$Zn(CH_3)_2$	Zn	$(2.0 \pm 0.6) \cdot 10^{-17}$	$< 10^{-18}$			38,90,135

[a]All measurements were made in the gas phase, and the values given represent an average of published constants. For simplicity, the precursors are listed in alphabetical order; for those precursors designed to yield a particular atom, ranking is generally according to the size of the molecule.

summarizes the photoabsorption cross sections for a variety of polyatomic precursors that are currently of interest for the photo-CVD of metal or semiconductor films (22). For most of the molecules in the table, an average of the cross-sectional values in the literature is given. In those instances where a considerable disparity (i.e., 1 order of magnitude or more) in the available values existed, the range of the reported constants is indicated. For convenience, most of the cross-sectional values given in Table 1 are those at the wavelengths of the rare gas–halide excimer lasers. (For several, the value of σ at 257 nm, the wavelength of the second harmonic of the Ar^+ ion laser line at 514.5 nm, is also given.) It should also be mentioned that, for a number of precursors, vapor pressures are not well known, which is a major source of uncertainty in σ.

To be a viable precursor for photo-CVD, a molecule must embody several essential characteristics. One is that it must be volatile; that is, its vapor pressure should exceed several tens of milliTorr at a temperature of $\leqslant 100°C$. Volatility is a requirement imposed by the need to transport the desired species to the deposition chamber. Most atomic elements of interest to microelectronics and manufacturing such as silicon, tungsten, and gallium have low vapor pressures ($< 10^{-5}$ Torr) at room temperature, and satisfying the aforementioned condition generally entails adding molecular ligands to the atom of interest [such as methyl groups, CH_3, to Si to form $Si(CH_3)_4$]. Of course, the resulting polyatomic molecule must also absorb optical radiation in a portion of the spectrum where efficient optical sources (lamps or lasers) exist. Other properties of precursors, which impact issues such as the purity and growth rate of the desired film, are equally critical and will be discussed as the chapter progresses.

While a variety of precursors have been explored for photo-assisted deposition, the most common are carbonyls, alkyls, or hydrides, which are not only (of necessity) volatile but also readily available. These parent molecules are of the form AB_m, where m is an integer and B is (respectively) CO, CH_3 (or C_2H_5 or iso-C_4H_9), or H. Halides have also been investigated but to a lesser degree.

Figures 4–7 (32, 33) show the UV and VUV absorption spectra for several of the most widely available hydride precursors for the Column IV atoms Si and Ge. For all of these, absorption reaches its peak value for wavelengths below 250 nm (or photon energies > 5 eV), which presents difficult decisions when choosing an optical source from the lamps and lasers that are commercially available (see Chapter 3, Section 3.3). As evidenced by the absorption spectra of the silanes in Figures 4a, 6, and 7, however, through the proper choice of precursor one does have modest control over the absorption spectrum. Note, in particular, the extension of the spectrum toward longer wavelengths as the size of the silane precursor rises. Of the existing lasers

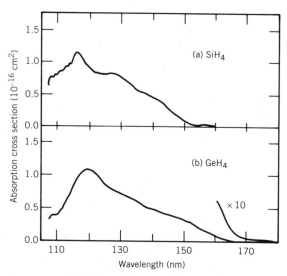

Figure 4. Absorption spectra for (a) monosilane, SiH_4, and (b) germane, GeH_4, in the VUV [reprinted by permission from Itoh et al. (*32*)]. As can be seen in Table 1, the photoabsorption cross section for both molecules at 200 nm is $< 10^{-19}\,cm^2$.

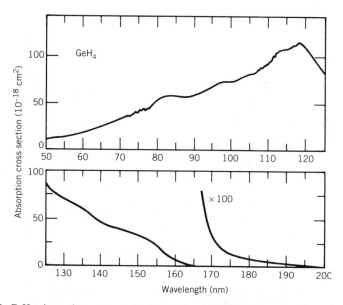

Figure 5. GeH_4 absorption spectrum in the VUV and XUV (extreme ultraviolet), 50–200 nm, showing an expanded view of the spectrum beyond 168 nm [reprinted by permission from Mitchell et al. (*33*)]. The cross section is given in units of $10^{-18}\,cm^2$.

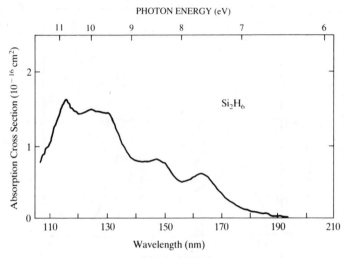

Figure 6. Disilane (Si_2H_6) absorption spectrum in the VUV ($110 \lesssim \lambda \leqslant 210$ nm) [after Itoh et al. (*32*), reprinted by permission]. Note the pronounced shift of this spectrum to longer wavelengths with respect to that for SiH_4 (see Figure 4a).

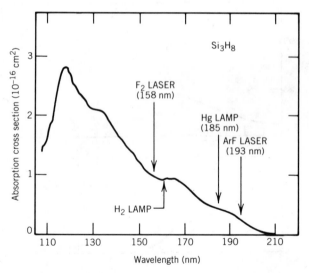

Figure 7. Absorption spectrum of trisilane (Si_3H_8), confirming that σ continues to increase at longer wavelengths [after Itoh et al. (*32*), reprinted by permission]. Also indicated are the wavelengths of the F_2 and ArF lasers, the 184.9 nm line emitted by low-pressure Hg lamps, and the 160 nm radiation produced by hydrogen lamps.

and lamps that are efficient, most (such as the rare gas–halide excimer lasers) operate in the spectral region above 200 nm, where the silanes are essentially transparent. Nevertheless, several sources emitting in the VUV (and the 150–200 nm range, in particular) have proven to be effective in photochemically depositing films from molecules such as the silanes that are characterized by large chemical bond strengths. Several of these sources are indicated in Figure 7 along with the absorption spectrum of trisilane, Si_3H_8.

Photoabsorption spectra of metal alkyl and hydride precursors for a variety of Group IIB, IIIB, VB, and VIB elements are given in Figures 8–12 (*34–38*). For these molecules, absorption beyond 200 nm is considerably stronger than was the case with the silanes and germanes, and photodecomposition with the longer wavelength excimer lasers (such as KrF, 248 nm) or high-pressure UV lamps becomes feasible. Note also that these spectra

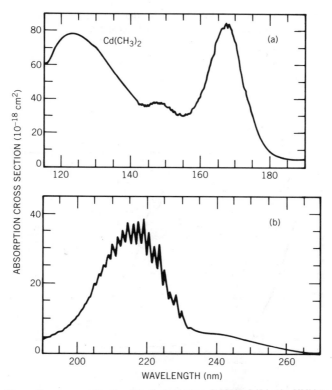

Figure 8. Absorption cross section for dimethylcadmium [$(CH_3)_2Cd$] in the VUV (top) and UV [adapted by permission from Suto et al. (*34*)]. The cross section is given in units of $10^{-18}\,cm^2$.

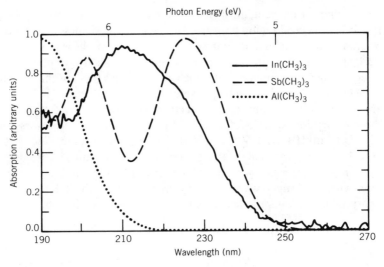

Figure 9. Trimethylaluminum, trimethylindium, and trimethylantimony absorption spectra in the UV (190–270 nm) (reprinted by permission from refs. *35* and *36*).

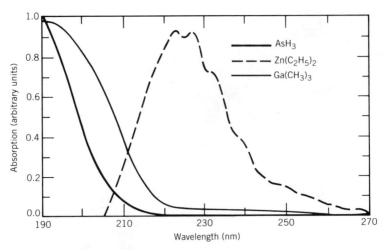

Figure 10. UV absorption spectra of arsine (AsH$_3$), diethylzinc, and trimethylgallium [reprinted from Hebner et al. (*36*), by permission].

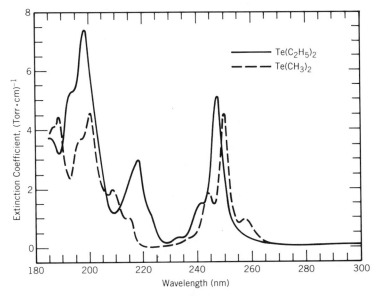

Figure 11. Absorption spectra of dimethyltelluride $[Te(CH_3)_2]$ and diethyltelluride $[Te(C_2H_5)_2]$ between 180 and 300 nm [after Zinck et al. (*37*)]. Similar spectra reported by Irvine et al. (*372*) display more discrete structure for $Te(CH_3)_2$ in the 205–220 nm region than that pictured here.

are essentially continua (displaying occasional structure) with absorption cross sections typically larger than $10^{-17}\,cm^2$. For the Group IB and VIII metals such as gold and platinum, respectively, the acetylacetonate precursors (known as "acacs") have proven useful, and Figure 13 (*39*) shows the UV absorption spectra for the platinum and iridium acetylacetonates. The Group VIA metals chromium, molybdenum, and tungsten form stable hexacarbonyls [i.e., $M(CO)_6$, where M is Cr, Mo, or W] and Figure 14 (*40*) shows the absorption spectrum of $Cr(CO)_6$ and $Fe(CO)_5$, which have arguably received the most attention of the three. The decomposition pathways for the chromium carbonyl will be discussed later in this chapter.

Most of the elements of interest to photodeposition are commercially available in the form of (at most) a few volatile precursors. This limited choice of precursors, particularly for metal atoms such as Au, Cr, and Ga, is an area that must be addressed for the potential of photo-CVD to be fully realized. The elemental and compound film growth techniques demonstrated to date have typically involved precursors that were developed previously for thermal (chemical vapor) deposition [MOCVD, CBE (chemical beam epitaxy), etc.], and future efforts will undoubtedly focus increasingly on

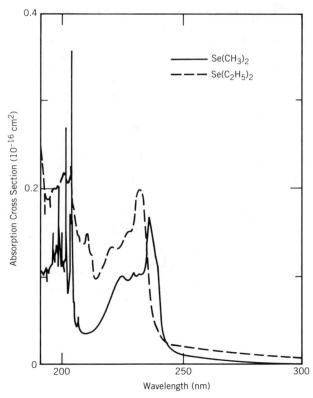

Figure 12. Dimethylselenide [$Se(CH_3)_2$] and diethylselenide [$Se(C_2H_5)_2$] absorption spectra in the UV [after Fujita et al. (*38*)].

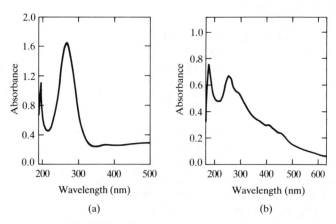

Figure 13. Vapor phase absorption spectra of the acetylacetonates (acacs) of (a) platinum and (b) iridium in the UV and visible [reprinted from Heidberg et al. (*39*), by permission].

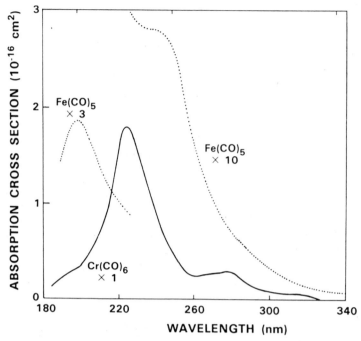

Figure 14. $Cr(CO)_6$ and $Fe(CO)_5$ absorption spectra [reprinted from Rothschild (*40*), by permission].

the synthesis and engineering of new precursors tailored specifically for photodeposition.

Although σ describes the degree to which power is absorbed by a given gas, it says nothing of the manner in which the energy is disposed. One of the factors contributing most to the complexity (and breadth!) of photochemical deposition is that the atomic and molecular fragments from a particular precursor are not only dependent upon the choice of parent molecule itself but are also a function of the wavelength of the source. Furthermore, although absorption cross sections are generally straightforward to measure, the products and their relative yields are difficult to ascertain, and for many of the photo-assisted deposition processes reported to date are not known. Several noteworthy exceptions to this statement are discussed later in this section, and research designed to fill this gap has intensified over the past few years. In summary, the amount of energy absorbed from the optical source by the gas (or adlayer) and the way in which it is distributed among the possible fragments (known as product channels or pathways) are both dependent upon wavelength. The remainder of this section briefly discusses the most common photochemical processes underlying photo-assisted deposition: photodissociation, photoionization, optically driven secondary (collisional) reactions, and multiphoton processes.

2.2.1. Photodissociation

Far and away the most common photochemical process invoked in photo-deposition is photodissociation, which involves the absorption of one or more photons by a molecule and ultimately results in the scission of the chemical bond. Photodissociation of a polyatomic molecule ABC can be expressed as

$$ABC + nh\nu \longrightarrow A + BC \qquad (5)$$

where A is the desired atom or molecular radical, BC represents the remaining molecular ligands, and n is an integer (generally, $n = 1$). As mentioned earlier, the identity of the photofragments and the distribution of energy (internal and translational) among the products are functions of wavelength. Also, photodissociation often results in the rupture of the weakest bond in the molecule, which may yield, for example,

$$ABC + h\nu \longrightarrow AB + C \qquad (6)$$

rather than the desired products of Reaction (5). Similarly, the fraction of the total absorbed photon energy that appears as translational energy (heat) of the fragments depends on the strength of the bond(s) that are broken, the number of bonds involved, and the laser wavelength.

Early photo-CVD experiments were largely empirical in the sense that little was known of the photoproducts that one could expect at a given wavelength. Consequently, only the most general features of the photochemistry could be inferred either from the known chemical and optical characteristics of analog molecules (such as CH_4 for SiH_4) or from the chemical composition of deposited films. Such a situation is not conducive to exploiting to the maximum the capabilities of gas phase photochemistry for film deposition, in general, and its ability to produce specific products, in particular. Recently, however, several studies have been reported that have investigated in detail the products of photodissociation for several precursors of interest to photo-CVD. A few of these will be reviewed here.

2.2.1.1. Group IIIB Alkyls

Not only are the Group IIIB alkyls such as trimethylaluminum [Al(CH)$_3$, TMA] or triethylgallium [Ga(C$_2$H$_5$)$_3$, TEG] widely used in the epitaxial growth of III–V semiconductor films (such as AlGaAs) by MOCVD, but they were also among the first precursors investigated for use in photo-CVD processes.

With a frequency-doubled, tunable dye laser and a time-of-flight mass spectrometer, Zhang, Beuermann, and Stuke (*41–43*) examined the products of the photodissociation of TMA and trimethylgallium (TMG) in the deep-UV. Since TMA is the most thoroughly studied precursor of the two and its photochemistry is representative (*41*) of that for other members of the Group IIIB alkyl family, TMG and trimethylindium (TMI), the discussion will be limited to its properties.

In the 190–255 nm spectral region, TMA in the vapor phase is photo-dissociated by the absorption of a single photon, giving rise to the absorption spectrum of Figure 15, which shows the 200–260 nm region in detail. The primary products are free Al atoms and $AlCH_3$ radicals and, as illustrated in Figures 16 and 17, the relative yields are strongly dependent upon laser

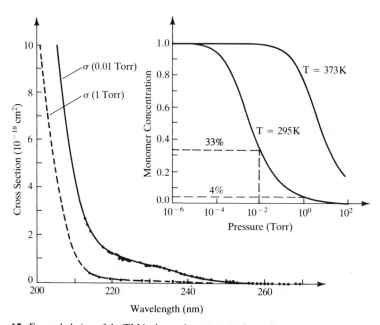

Figure 15. Expanded view of the TMA absorption spectrum between 200 and 260 nm [reprinted by permission from Beuermann and Stuke (*41*)]. As shown by the inset, the cross section is dependent upon pressure because of the rising concentration of the TMA dimer ($Al_2(CH_3)_6$) at higher pressures. At room temperature, the TMA monomer ($Al(CH_3)_3$) relative number density falls from 33% at a pressure of 10^{-2} Torr to 4% at 1 Torr of total pressure.

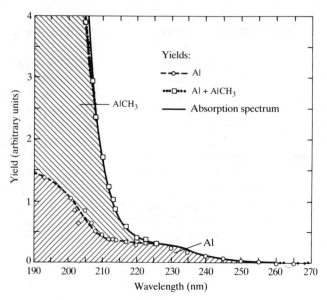

Figure 16. Relative yields of Al and AlCH$_3$ resulting from the photodissociation of TMA in the 190–270 nm spectral region [reprinted by permission from Beuermann and Stuke (*41*)]. The absorption spectrum for TMA is represented by the solid curve. Note that for wavelengths greater than $\sim$ 220 nm, Al is the predominant product.

wavelength (*41*). For wavelengths below 255 nm, Al atoms are produced by the two-step process (*41, 44*):

$$Al(CH_3)_3 + h\nu \ (\geqslant 4.9\,eV) \longrightarrow Al(CH_3)_2^\dagger + CH_3 \tag{7}$$

$$Al(CH_3)_2^\dagger \longrightarrow Al + .C_2H_6 \tag{8}$$

where the dagger denotes electronic (i.e., internal) excitation of the dimethyl-aluminum radical. Postulating the production of C$_2$H$_6$ appears to be necessary to explain the formation of free Al atoms for photon energies as low as 5.0 eV despite the fact that the average Al—CH$_3$ bond energy in TMA is roughly 2.9 eV. Since the energy expected to be necessary to break all three Al—CH$_3$ bonds is 8.7 eV, the discrepancy of 3.7 eV requires the production of the stable product C$_2$H$_6$ in Eq. (8).

The threshold for AlCH$_3$ production was observed by Beuermann and Stuke (*41*) to be $\sim$ 230 nm. For this fragment, the production process appears to be

$$Al(CH_3)_3 + h\nu \ (\geqslant 5.4\,eV) \longrightarrow AlCH_3 + 2CH_3 \tag{9}$$

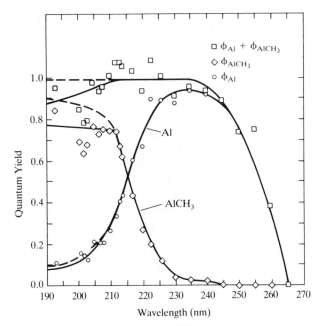

Figure 17. Absolute quantum yields for Al and $AlCH_3$ in the photodissociation of TMA (*41*). The dominance of atomic aluminum as a product at longer wavelengths is again clearly evident.

Stipulating two methyl radicals (rather than one ethane molecule) as the products in Eq. (9) is a direct consequence of the observation that the appearance of $AlCH_3$ occurs at higher photon energies than does Al production. If C_2H_6 were produced, $AlCH_3$ fragments would appear at *lower* energies than would Al atoms, but such is not the case. TMA, TMG, and TMI are apparently all photodissociated in the same manner (*41*).

For $\lambda > 220$ nm, it was also found that only the TMA monomer $[Al(CH_3)_3]$ absorbs, and the dimer $[Al_2(CH_3)_6]$ contribution rises rapidly at shorter wavelengths. The results of refs. *41–44* explain the presence of objectionable levels of carbon in Al films deposited by photodissociating TMA at 193 nm (although adlayer photoprocesses must also be considered and are discussed in Section 2.3, below), suggesting that a future course would be to use a source in the $\lambda \geqslant 220$ nm region—perhaps the KrCl laser at 222 nm. As will be discussed in Chapter 4, experiments in which TMA was photodissociated at 248 nm (KrF laser) to deposit Al films having lower carbon impurity concentrations support the conclusions of Beuermann, Zhang, and Stuke (*41–43*).

Similar processes appear to hold sway in other Al-alkyls as well. Brum et al. (45) photodissociated triethylaluminum [Al(C$_2$H$_5$)$_3$, TEA] at 193 nm and, as a result of monitoring the free hydrogen produced, postulated that the process is initiated by breaking an Al—C bond, which yields an ethyl radical. The overall process was shown to be dominated by the absorption of a single photon per TEA molecule.

2.2.1.2.   Group VB Hydrides

The Column VB hydrides NH$_3$ (ammonia), PH$_3$ (phosphine), AsH$_3$ (arsine), and SbH$_3$ (stibine) are also frequently employed in the MOCVD and CBE growth of III–V compound semiconductor films. All have been instrumental in the development of a host of optoelectronic devices, including quantum well structures for both sources and detectors. Despite the toxicity of these molecules, the quality of compound films grown from the hydride precursors is superior to that from other feedstock gases such as the alkyls. Consequently, photodeposition as well as conventional growth techniques have focused on these polyatomics.

All of the Group VB hydrides absorb strongly in the VUV, with absorption maxima occurring in the 180–200 nm region (29, 40) and, except for ammonia, their absorption spectra are continua. The NH$_3$ spectrum is oscillatory in the 180–200 nm region, and the 193 nm cross section is $1.2–1.7 \times 10^{-17}$ cm^2 (46–48). Photodissociation of the molecule at these photon energies results in the removal of a single hydrogen atom. For phosphine, for example (40, 49),

$$PH_3 + h\nu \longrightarrow \begin{cases} PH_2\,[\tilde{A}(^2A_1)] + H \\ PH_2\,[\tilde{X}(^2B_1)] + H \end{cases} \tag{10}$$

where only $\sim 1\%$ of the PH$_2$ radicals are produced in the excited ($\tilde{A}$) state; most are formed directly in the ground state. Similar statements could be made for the other hydrides—for NH$_3$, the yield into the NH$_2$ ($\tilde{A}$) state is 2.5%, with the remainder produced once again in the $\tilde{X}$ ground state (50).

2.2.1.3.   Metal Carbonyls

The carbonyls of chromium, tungsten, iron, molybdenum, manganese, and nickel absorb strongly below 300 nm, and in most instances the dissociation of these polyatomics proceeds by the sequential elimination of carbonyl ligands. Although the energy required to remove one or two CO ligands is 1.6 and 3.3 eV, respectively (51), the photodissociation of chromium hexacarbonyl, for example, in the 351–355 nm region, proceeds predominantly by the process (52–55):

$$Cr(CO)_6 + h\nu\,(\simeq 3.5\,eV) \longrightarrow Cr(CO)_5 + CO \tag{11}$$

At 248 nm, while two photons have sufficient energy (10 eV) to remove *all* of the carbonyl ligands (55), a single 5-eV photon leads to the removal of two carbonyls by the two-step sequence (55–58):

$$Cr(CO)_6 + h\nu\,(5\,eV) \longrightarrow [Cr(CO)_5]^* + CO \longrightarrow Cr(CO)_4 + 2CO \tag{12}$$

where the asterisk indicates that the pentacarbonyl species is unstable. A small amount of $Cr(CO)_3$ is also produced when $Cr(CO)_6$ is photodissociated with a KrF laser (248 nm) (56). When $Cr(CO)_6$ is photolyzed at 193 nm (ArF laser), $Cr(CO)_4$, $Cr(CO)_3$, and $Cr(CO)_2$ are generated (56, 59). At laser intensities exceeding several hundred kilowatts per square centimeter, electronically excited metal atoms are produced as a result of two-photon absorption by the parent molecule (60). Similar observations have recently been reported by Radloff et al. (61).

2.2.2. Photoionization

Of the techniques developed to date to grow semiconductor or metal films from the vapor phase, virtually all deal with chemical reactants and products that are electrically neutral. However, the selective photoionization of reactants presents several tantalizing advantages for photo-CVD, insofar as process control, impurity rejection, and chemical specificity, in particular, are concerned. One attractive aspect is that, since the product is charged, selective area deposition by electrostatic control is feasible. Also, as the wavelength of the optical source is reduced in moving from the visible to the UV, VUV, or beyond, accessing exit channels having high-energy thresholds (such as photoionization) becomes increasingly likely. A drawback of this approach is that deposition rates are typically quite low ($< 500\,\text{Å}\cdot\text{h}^{-1}$).

2.2.2.1. *Ion Pair Production in Metal-Halides*

Upon absorbing a single photon of sufficiently high energy, the entire molecule AB can be photoionized:

$$AB + h\nu \longrightarrow (AB)^+ + e^- \tag{13}$$

or the molecule may be dissociated in the process, yielding a single, positively charged ion and an electron:

$$AB + h\nu \longrightarrow A^+ + B + e^- \tag{14}$$

or only heavy ions:

$$AB + h\nu \longrightarrow A^+ + B^- \tag{15}$$

The first of these process [Reaction (13)] is generally not useful in photo-CVD since the precursor ligands are usually required *only* as a vehicle for raising the volatility of the parent. Once the precursor is present at the substrate, the ligands are no longer necessary (and are often a nuisance, as they frequently contain undesirable atoms such as carbon) and the molecule must be dismantled so as to gain access to the desired atom. Occasionally, however, photoionization of the entire precursor molecule at short (i.e., VUV) wavelengths is an attractive deposition process, particularly for hydrogenated films (see Chapter 3, Section 3.1) (*62*).

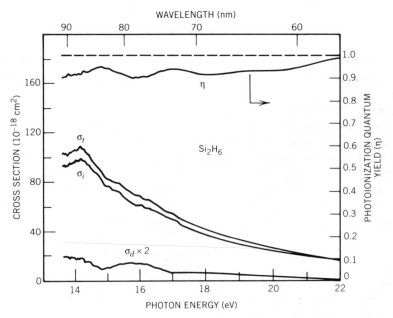

Figure 18. Photoabsorption and photoionization spectra of Si_2H_6 in the VUV and XUV [after Kameta et al. (*63*)]. The total photoabsorption cross section for the molecule (expressed in $10^{-18} cm^2$) is indicated by σ_t. The apportionment of the absorbed energy between photodissociation and photoionization of the molecule is indicated by the cross sections σ_d and σ_i, respectively.

The introduction of new product channels that occur as the wavelength of the optical source is decreased was noted earlier. Recent data (63) obtained for Si_2H_6 provide an illustration of the interaction between neutral fragmentation (photodissociation) and photoionization. Recall from Figure 6 that disilane absorbs weakly for wavelengths longer than roughly 190 nm. In the long-wavelength portion of the absorption spectrum, *primarily* neutral products (such as SiH_3 radicals) are produced but, as shown in Figure 18 (63), the role of ionization increases dramatically as the photon energy rises. The apportionment of the absorbed energy between the generation of neutral and ionized products is indicated by the cross sections σ_d and σ_i, respectively. In this wavelength region, photoionization of the molecule is clearly the dominant channel and dissociation into neutral species accounts for less than 10% of the products.

An excellent example of the reaction described by Eq. (15), however, is the *dissociative* photoionization of the Group IIIB iodides as well as other metal-halides. In the early 1930s, Terenin and Popov (64, 65) found that certain metal-halide molecules, when irradiated with photons lying within a narrow energy range, undergo dissociative photoionization, resulting in the production of only charged fragments—a metal cation and a halogen anion. As an example, consider the diatomic metal-halide thallium iodide (TlI), whose absorption spectrum (66) in the 190–210 nm region is shown in Figure 19 (66). Measurements by Terenin and Popov (64, 65) and Berkowitz and Chupka (67) reveal that a significant fraction of the optical energy absorbed in this spectral region results in the production of Tl^+-I^- ion pairs. The relative efficiency for the production of Tl^+ by the process

$$TlI + h\nu \longrightarrow Tl^+ + I^- \tag{16}$$

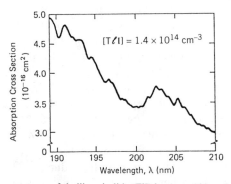

Figure 19. Absorption spectrum of thallium iodide (TlI) between 190 and 210 nm [after Geohegan and Eden (66)]. Note that the ordinate gives the absorption cross section in units of $Å^2$ ($= 10^{-16} cm^2$).

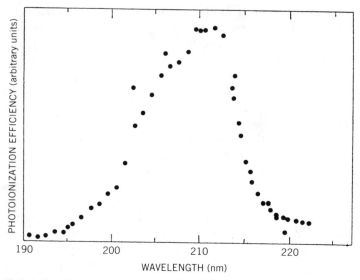

Figure 20. Relative efficiency for the production of Tl^+ ions by the dissociative process: $TlI + hv \rightarrow Tl^+ + I^-$ in the 190–220 nm spectral interval [reprinted by permission from Berkowitz and Chupka (67)]. Note that all of the products are charged. Also, analogous processes occur in other metal-halides.

is illustrated in Figure 20. Note that peak ion pair production occurs at ~210 nm . The three most appealing features of dissociative photoionization of metal halides insofar as photochemical vapor deposition is concerned are: (1) by the careful choice of the precursor and laser wavelength, only one positively charged ion is produced (that of the metal atom) and the undesired species (halogen atom) is oppositely charged—consequently, with the aid of an electric field, one can minimize the incorporation of contaminants into the resulting film (68); (2) because of the large electronegativities of the halogen atoms (3.06 eV for iodine), the energy threshold for $Tl^+—I^-$ production, for example, is considerably lower than that for the formation of $Tl^+ + I + e^-$; and (3) lower temperature processing is possible since the metal halides frequently have higher vapor pressures than those for the metal itself and the substrate can be physically separated from the ion production region. The second of these advantages is depicted in Figures 21 and 22, which show the onset for the production of $Tl^+ + I + e^-$ (Eq. 14) and $(TlI)^+$ (Eq. 13), respectively, for optical source wavelengths below 150 nm. Note that the energy thresholds for the formation of both products are nearly equal, which severely restricts the photochemical selectivity that is available. Furthermore, no commercially available lasers operate in this spectral region whereas, at longer wavelengths ($\lambda = 158, 193, 248,\ldots$ nm), several excellent pulsed and

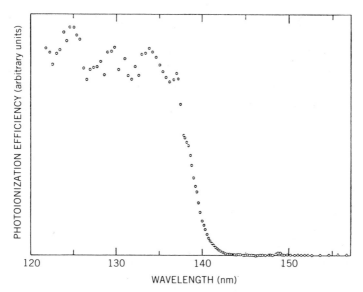

Figure 21. Relative efficiency for the photoionization of TlI to yield $Tl^+ + I + e^-$ in the VUV ($120 \leqslant \lambda \leqslant 150$ nm) [reprinted by permission from Berkowitz and Chupka (67)].

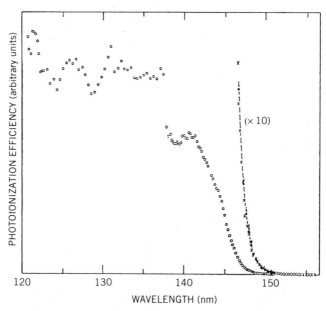

Figure 22. Relative production efficiency for $(TlI)^+$ in the spectral region below 155 nm [reprinted by permission from Berkowitz and Chupka (67)]. Note that the energy thresholds for the production of $(TlI)^+$ and $Tl^+ + I + e^-$ (see Figure 21) are quite similar.

continuous wave (CW) laser choices exist. Similar considerations are, of course, valid for other molecules such as TMA. Calloway et al. (69) have shown that thin Al films can be deposited by photoionizing TMA with rare gas lamp resonance radiation ($\lambda \simeq 110$–180 nm; see Figure 43 in Chapter 3, Section 3.3.1). At these short wavelengths, however, the entire molecule is

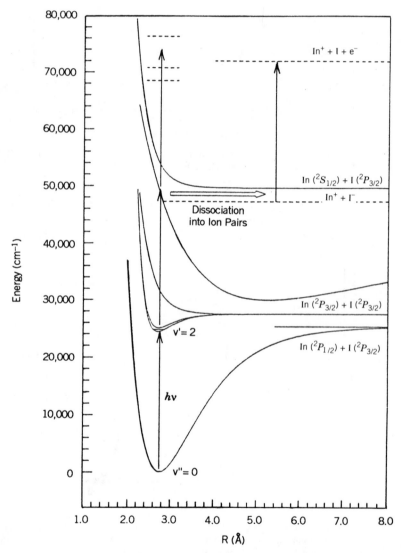

Figure 23. Partial energy level diagram for indium monoiodide illustrating the production of In^+–I^- pairs by one- or two-photon processes [after King (70)].

apparently ionized, leading to incorporation of carbon (and hydrogen) into the film.

Ion pair production processes analogous to reactions (15) and (16) occur in other metal-halides such as indium monoiodide (InI), and Figure 23 is a partial energy level diagram for the molecule (70). ArF laser photons ($hv = 6.4\,eV$) have sufficient energy to directly produce In^+-I^- ion pairs, and this process competes with the formation of electronically excited, neutral In atoms and a ground state I atom (71). [This process is dominant for TlI in the region between ~ 150 and $200\,nm$ (Figures 20–22) and competes with Tl^+-I^- production.] As illustrated qualitatively in Figure 23, In^+ ions can also be produced by the absorption of two visible photons (70). At longer wavelengths ($\lambda \gtrsim 360\,nm$), the absorption of a single photon yields two ground state atoms.

2.2.2.2. Group VB Hydrides

Koplitz et al. (72) have noted that "given its obvious industrial and economic significance, there is surprisingly little detailed knowledge of the photophysics and photochemistry of AsH_3," and the same may be said for the other Group VB hydrides. In a clever series of experiments, Koplitz, Xu, and Wittig (72)

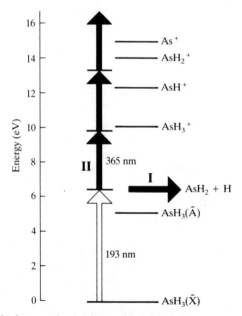

Figure 24. Generalized energy level diagram for AsH_3 showing the switch between photo-dissociation (denoted pathway **I**) and dissociative photoionization (denoted **II**) that is induced by simultaneously irradiating the molecule with 6.4 and $\sim 3.4\,eV$ photons [reprinted from Koplitz et al. (72), by permission].

demonstrated that the photochemistry of AsH_3 can be readily switched from predominantly neutral to ionic products with a "two-color" laser approach. As noted earlier, the absorption of a single 193-nm photon by AsH_3 predominantly produces AsH_2 radicals in the $\tilde{X}$ (ground) or $\tilde{A}$ (excited) electronic states. Upon adding $\lambda \simeq 365$ nm (3.4 eV) photons concurrently with the ArF radiation, strong ionization of the molecule and the formation of AsH^+, in particular, is observed. Consequently, the AsH_3 fragments can be readily manipulated with the aid of the UV radiation, which acts in concert with 6.4 eV photons to photoionize $AsH_3(\tilde{A})$ molecules. This scheme can be expressed as

$$AsH_3 + h\nu_1\,(6.4\,eV) \longrightarrow AsH_2\,(\tilde{X}, \tilde{A}) + H \qquad (17)$$

$$AsH_3 + h\nu_1 + h\nu_2\,(3.4\,eV) \longrightarrow AsH_x^+ + e^- + H \qquad (x \leqslant 3) \qquad (18)$$

and is shown schematically in Figure 24.

Similar approaches are almost certainly available for other precursors of interest to photo-CVD, but further fundamental studies of the photochemistry of these molecules will be necessary to assess the full scope of the chemical pathways and subsequent process engineering that is possible.

2.2.3. Optically Driven Secondary Reactions

2.2.3.1. Sensitizers

Often, a mismatch exists between the wavelengths available from existing (and efficient) UV and VUV lasers (and lamps) and the spectral regions in which volatile, high-purity precursors absorb. A more subtle difficulty arises if, as discussed previously, the photodecomposition of a given precursor at a particular optical source wavelength *does* occur rapidly but yields undesirable products. In such situations, intermediate species, known as *photosensitizers*, may occasionally be employed to act as a bridge between the optical source and the precursor molecule—in the sense that the photosensitizer species must absorb strongly at the optical source wavelength and efficiently transfer its internal energy to the polyatomic precursor molecule.

Atomic mercury is the oldest and most thoroughly studied photosensitizer, as the decomposition of silane by mercury photosensitization was reported more than a half century ago (2). Since that time the process has been investigated by a large number of groups (136–139).

Mercury is an effective photosensitizer for the Column IV hydrides in that it couples Hg resonance lamps emitting at 253.7 (and 184.9) nm with precursors such as SiH_4 and Si_2H_6, which are virtually transparent for wavelengths beyond 160 and 200 nm, respectively (see Figures 4 and 6). Consequently, in the absence of a photosensitizer, decomposing precursors such as SiH_4 by single-photon processes precludes the use of the most efficient, commercially available optical sources.

Central to Hg photosensitization is the $6p^3P_1$ electronic excited state of the atom, which lies ~ 4.9 eV above the ground state. Mercury $^3P_1$ atoms, produced by 254-nm radiation from a low-pressure resonance lamp, are well suited for collisionally exciting (and dissociating) various polyatomic molecules. In fact, the application of such reactions to thin film deposition is a subset of the much larger field of Hg-sensitized reactions. A more detailed discussion of the photochemistry of the $Hg(6p^3P_1)$ species can be found in Okabe (29).

The collisional processes that are critical to the photochemical growth of amorphous hydrogenated silicon (a-Si:H) films from the photosensitized decomposition of SiH_4 are (139)

$$Hg + h\nu \longrightarrow Hg^* \tag{19}$$

$$Hg^* + SiH_4 \longrightarrow Hg + SiH_3 + H \tag{20}$$

$$H + SiH_4 \longrightarrow H_2 + SiH_3 \tag{21}$$

$$SiH_3 + SiH_3 \longrightarrow SiH_2 + SiH_4 \tag{22}$$

$$SiH_3 + SiH_3 \longrightarrow SiH_3SiH + H_2 \tag{23}$$

where the asterisk denotes the $6p^3P_1$ excited state of mercury. Although the actual production of a-Si:H films appears to occur when SiH_2SiH_3 or SiH_3SiH radicals impinge on the substrate (139), efforts to unravel the complicated gas phase chemistry of this photochemical system are continuing. Nevertheless, the chemical versatility of Hg photosensitization, combined with the economy and simplicity of Hg resonance lamps, makes this an attractive approach to the deposition of the Column IV elemental films, in particular.

Despite the simplicity of Hg photosensitization, the addition of Hg vapor to the photochemical reactor (and incorporation of Hg into the resulting films) is often undesirable and it is essential that photochemical intermediates other than Hg* be explored. It has been shown (140) recently, for example, that photochemically generated hydrogen atoms accelerate the decomposition of germane (GeH_4), enabling epitaxial Ge films to be grown on GaAs at substrate temperatures as low as 300°C. Free H atoms are produced by photodissociating ammonia in the VUV (50, 141):

$$NH_3 + h\nu \xrightarrow{a} NH_2(\tilde{X}^2B_1) + H \tag{24}$$

$$\xrightarrow{b} NH_2(\tilde{A}^2A_1) + H \tag{25}$$

where the relative yields a and b are 97.5% and 2.5%, respectively, at a

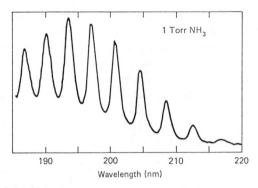

Figure 25. Absorption spectrum of NH_3 in the 185–220 nm region. The absorption cross section at 193 nm (the wavelength of the ArF excimer laser) is $\sim 1 \times 10^{-17}$ cm^2.

wavelength of 193 nm (50). Photodissociation of GeH_4 at 193 nm *also* yields hydrogen atoms, but the absorption cross sections for GeH_4 and NH_3 at this wavelength differ by 3 orders of magnitude (see Figure 25). Once produced, H atoms decompose the hydride by abstraction collisions:

$$H + GeH_4 \longrightarrow GeH_3 + H_2 \tag{26}$$

which is analogous to Reaction (21) for monosilane. Consequently, the laser serves to artificially increase the H atom concentration above the substrate; the photochemical production of H in essence allows one to break into the kinetics chain of Eqs. (19)–(23) without the need for atomic Hg to be present in the reactor. Trace amounts of NH_3 added to a He/GeH_4 gas flow stream have been observed to increase the growth rate of epitaxial Ge films by more than 1 order of magnitude (*140*).

It should be noted that Reactions (24)–(26) represent the optical version of *h*ydrogen-*r*adical-assisted *c*hemical *v*apor *d*eposition (HRCVD), which is based on H atoms produced in plasmas and has, for several years, been applied to the growth of semiconductor films (*142–145*). Two advantages of the optical approach are its ability to produce the H atoms *in situ* (in the immediate vicinity of the substrate) and its applicability to a variety of hydrogen-bearing precursors. With regard to the latter, H_2S appears to be equally acceptable to NH_3 as a hydrogen donor (*146, 147*).

2.2.3.2. *Bimolecular Collisional Formation of Compounds*

Compound film growth by photochemical vapor deposition is, at present, a class of complex and poorly understood processes. The deposition of the compound material $A_x B_y$ (where x and y are often *not* integers) usually requires molecular precursors for both elements A and B and is similar to photosensitized deposition in that generally only one of the two precursors absorbs strongly at the optical source wavelength. Upon absorbing a photon, this precursor is either excited to a relatively stable excited state or is dissociated. In either case, the products formed by irradiating the molecule are more chemically reactive than are the ground state species and the ensuing collisions of these photofragments with the remaining precursor eventually culminates in the production of the desired film.

An example of this process is the photochemical sequence responsible for the deposition of the insulator films Si_3N_4 and SiO_2. In both cases, the silicon precursor is a silane (SiH_4, Si_2H_6, or Si_3H_8), all of which, as noted earlier, are essentially transparent (nonabsorbing) in the UV. Interaction of the optical source occurs primarily with the N or O precursor whose dissociation fragments chemically react with the silane and the surface to form the desired film. In the photodeposition of SiO_2, the oxygen precursor is frequently N_2O, which, when photodissociated at 193 nm, yields

$$N_2O + h\nu\,(\sim 6.5\,\text{eV}) \longrightarrow N_2 + O^*\,(^1P) \tag{27}$$

Following relaxation of the excited $O\,(^1P)$ species to the ground (^3P) state, compound formation apparently proceeds by the reaction (148)

$$SiH_4 + 4O\,(^3P) \longrightarrow SiO_2 + 2H_2O \tag{28}$$

The photochemistry of Si_3N_4 deposition is considerably more complicated, since it entails the photodissociation of SiH_4 (or Si_2H_6)/NH_3 mixtures in the VUV and generally at 193 nm. Two of the early steps in the reaction chain are expected to be

$$NH_3 + h\nu \longrightarrow NH_2\,(\tilde{X}, \tilde{A}) + H \tag{24, 25}$$

$$H + SiH_4 \longrightarrow SiH_3 + H_2 \tag{21}$$

with the final deposit-producing reactions involving the interaction of laser-generated radicals (such as amidogen, NH_2) with SiH_4. The understanding of such complex, multistep reactions is still in its infancy, and considerable theoretical and experimental study will be required in order to grasp the process at a level where it can be optimized and generalized to the deposition of other films.

2.2.4. Multiphoton Processes

The photoreactions discussed to this point have generally assumed the absorption of a single photon per precursor molecule as a necessary condition for triggering the photodeposition process of interest. As we have seen, however, several parent molecules of interest (such as SiH_4) are "transparent" in spectral regions where efficient optical sources presently exist. While transparency is a notion that is a result of a small *single*-photon absorption cross section, σ, *multiphoton* processes often become important at higher optical intensities and can alter the situation entirely. In other words, most absorption spectra (including those presented earlier in this chapter) were recorded at low optical probing intensities and therefore generally reveal only those spectral regions in which the molecule in question absorbs a single photon. As the optical intensity is increased, however, molecules will also simultaneously absorb two or more photons, thus giving access to an entirely new array of possible products. Prior to the advent of lasers, such multiphoton processes were rare (because of the limited intensities available from lamps), but the development of various tunable, high-peak-power lasers (i.e., $> 10^6$ W) has established multiphoton fragmentation and ionization as an addition tool for the photochemist.

Recalling the generalized photodissociation process,

$$ABC + nh\nu \xrightarrow{\sigma^{(n)}} \text{PRODUCTS} \tag{5}$$

where n is again an integer, we denote the nth-order photoabsorption cross section as $\sigma^{(n)}$ and the rate $\mathscr{R}$ at which ABC molecules are photodissociated (or photoionized) is given by perturbation theory as

$$\mathscr{R} = \sigma^{(n)}\left(\frac{I}{h\nu}\right)^n \tag{29}$$

where I is the optical intensity (expressed in $W \cdot cm^{-2}$) and $\sigma^{(n)}$ is expressed in units of $cm^{2n} \cdot (s)^{n-1}$ (i.e., cm^2 for $n = 1$; $cm^4 \cdot s$ for $n = 2$; $cm^6 \cdot s^2$ for $n = 3; \dots$). Note the strong nonlinearity of the process for $n > 1$ (i.e., the absorption of more than one photon by the precursor), which favors large laser intensities. Figure 26 qualitatively illustrates the absorption of three photons of equal energies by an atom or molecule, resulting in the excitation of the species from an initial state i to a higher lying, terminal energy level, denoted f. It is not necessary for real states of the atom or molecule to lie in the energy region between i and f. However, if the photon energies and the species under study are chosen such that, as shown in Figure 26b, the photon energies

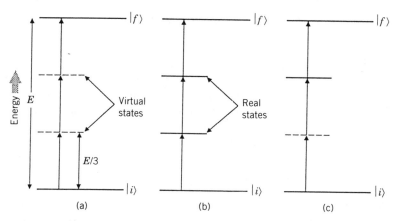

Figure 26. Generalized diagram for a molecule (M) illustrating three-photon absorption by (a) nonresonant, (b) resonant, and (c) a combination of resonant and nonresonant processes.

coincide with differences between real energy states of the molecule (or atom), then the multiphoton transition rate will be enhanced greatly over the *nonresonant* value.

As an example, silane absorbs weakly at 193 nm ($hv \simeq 6.4\,\text{eV}$) owing to a single-photon cross section at that wavelength of $\sigma^{(1)} = 1.1 \times 10^{-21}\,\text{cm}^2$ (*82*). However, the *two*-photon cross section, $\sigma^{(2)}$, at the same wavelength is $(6 \pm 2) \times 10^{-44}\,\text{cm}^4 \cdot \text{s}$ (*149*), and since the ratio of the single-photon absorption rate to that for two photons is $(\sigma^{(1)}/\sigma^{(2)})\,hv/I$, then the rates for these two processes are equal for laser intensities of only $\sim 19\,\text{kW} \cdot \text{cm}^{-2}$ (or $0.2\,\text{mJ} \cdot \text{cm}^{-2}$ for 10-ns laser pulses). This intensity is easily exceeded with commercially available excimer lasers, which generally produce unfocused intensities of $> 4\,\text{MW} \cdot \text{cm}^{-2}$ per pulse. Consequently, two(or more)-photon processes can dominate single-photon dissociation of a molecule and, in such cases, conventional absorption spectra are of little value.

Another example of multiphoton absorption in precursors of interest to photoassisted deposition includes

$$\text{GeH}_4 + 2hv\,(\lambda = 248\,\text{nm}) \xrightarrow{\sigma^{(2)}} \text{GeH}_2 + 2\text{H} \tag{30}$$

where $\sigma^{(2)}$ has been estimated to be $> 4 \times 10^{-47}\,\text{cm}^4 \cdot \text{s}$ (or $> 5 \times 10^{-29}\,\text{cm}^4/\text{W}$ for 5 eV photons) (*101*). For other precursors such as the metal carbonyls, the absorption of two or more photons provides sufficient energy to completely liberate a metal atom (*60*). Tyndall et al. (*150*) have shown that KrF laser (248-nm) irradiation of various chromium-bearing molecules (such as benzene–Cr(CO)$_3$) results in the two-photon production of electronically excited Cr atoms. Similar observations have been reported by Radloff and co-workers (61) for Mo(CO)$_6$ at 248 nm.

Rarely are higher order ($n > 2$) multiphoton absorption processes of interest for photo-assisted deposition, but Hackett, John, and Mitchell (151–153) have demonstrated that multiphoton ionization spectrum of TMA in the *visible* region is characteristic of that for the aluminum atom. That is, the absorption of five blue ($\lambda \sim 445$ nm) photons will completely remove all of the methyl ligands and yield an Al$^+$ ion:

$$Al(CH_3)_3 + 5hv\,(\sim 2.8\,eV) \longrightarrow Al^+ + e^- + 3CH_3 \qquad (31)$$

Owing to the large optical intensities required, multiphoton processes similar to Eqs. (30) and (31) are accessible only with pulsed laser sources.

In summary, whether the pivotal photochemical process is photodissociation, photoionization, or photosensitization, the role of the optical source in photochemical vapor deposition is to transform the chemical environment in the vicinity of the substrate from quiescence to a reactive one but to do so in a manner that discriminates against undesired products. Through a careful selection of the precursor(s) and optical source wavelength, photochemical processes are capable of providing product selectivity that is not generally available through thermally driven deposition (pyrolysis) alone.

2.3.   OPTICAL CHARACTERISTICS OF MOLECULAR ADLAYERS

Photochemical interactions in adlayers of molecules at a surface underlie many photochemical deposition processes and are particularly useful in those applications for which spatial resolution is an important consideration. In the gas phase, the lateral diffusion of photoproducts imposes a limit on the smallest feature sizes obtainable by photo-assisted film deposition and, while this can be suppressed to some extent by controlling the identity and pressure of the background gas (154), the restricted mobility of atomic and molecular species in an adlayer is much more conductive to depositing patterned films.

The proximity of a surface introduces a new dimension to photo-CVD that is not usually a factor in traditional gas phase photochemistry, and yet surface-mediated reactions are capable of significantly altering the chemical dynamics of the system. Furthermore, the optical and chemical properties of adsorbed layers of polyatomic molecules can differ considerably from those in the gas phase. Qualitatively, adlayers can be categorized as *chemisorbed* or *physisorbed*. For the former, the interaction of the adlayer with the surface is comparable to the precursor's intramolecular bond strengths (40). Physisorbed adlayers, in contrast, often consist of 10 or more monolayers of molecules (or atoms) that interact weakly with the surface, and their absorption spectra are generally similar to those for the gas phase species.

Although the characteristics of adsorbed layers of a variety of molecules have been studied extensively, considerably less is known regarding adlayers of the polyatomic molecules that are suitable precursors for photo-CVD. Of the few precursors that have been studied, the Column IIIB alkyls are the most thoroughly characterized and will be discussed here. Figure 27 compares the gas phase absorption spectrum of $Ga(CH_3)_3$ (TMG) with the physisorption and chemisorption spectra recorded by Sasaki et al. (155) for TMG adsorbed onto SiO_2. The chemisorbed spectrum appeared once TMG was introduced onto the surface and, because of its strong interaction with the surface, remained even after the optical cell was evacuated, which had

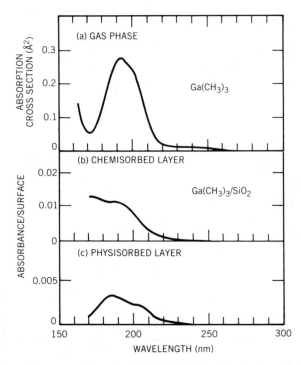

Figure 27. Comparison of the absorption spectrum for (a) gas phase $Ga(CH_3)_3$ with the spectra for (b) chemisorbed and (c) physisorbed adlayers of TMG (reprinted by permission from ref. 155). Spectra b and c were acquired for a TMG partial pressure of 0.22 Torr.

the effect of removing the gas phase molecules and the weakly bound physisorbed species. Note that the chemisorbed and physisorbed spectra are both broader than their gas phase counterpart and that, relative to the gas phase profile, they peak at shorter wavelengths. Nevertheless, the absorption spectrum of the physisorbed layer closely resembles the gas phase spectrum.

Similar results have been obtained for dimethylcadmium (DMCd) and TMA. Shaw et al. (*156*) measured the absorption spectra of gas phase and chemisorbed DMCd (on SiO_2) and, as illustrated in Figure 28, the spectra are similar although the chemisorption spectral peaks are broad continua because of the surface interaction. Also, the absorption cross section for chemisorbed DMCd was estimated to be $2 \times 10^{-17} \, cm^2$—a value that is roughly a factor of 2 larger than the gas phase value (see Table 1). Earlier studies of Ehrlich and Osgood (*157*) on DMCd and TMA and the experiments of Rytz-Froidevaux et al. (*158*) and Chen and Osgood (*159*) on DMCd reveal an elongation of the chemisorbed layer absorption spectrum toward the red. Spectral features not present in the gas phase spectrum are also often

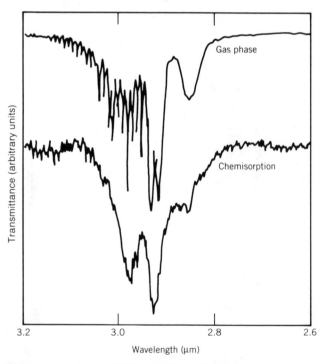

Figure 28. IR absorption spectra for DMCd $[Cd(CH_3)_2]$ in the gas phase and chemisorbed onto SiO_2 (after ref. *156*, by permission).

observed for the chemisorbed species. For TMA adlayers on quartz substrates, for example, undulations in the chemisorbed spectrum were reported in Ref. *157* with local maxima lying at ~ 200 and 260 nm. Another significant attribute of chemisorbed adlayers insofar as photodeposition is concerned is that, as cited earlier for DMCd, other studies indicate that the absorption cross section of the adlayer is generally larger than that for the vapor (*155, 157, 159*).

2.4. SURFACE KINETICS

Regardless of whether films are photodeposited with direct optical illumination of the surface or not, gas phase chemistry describes only a portion of the overall process and surface reactions will invariably play a significant or dominant role. It is the interactions of precursor molecules or their photofragments with the surface that determines the growth rate, purity, and structure of the resulting films. Photodeposition is an umbrella term for a complex group of processes that are ongoing at any given time but can be broken down into several basic steps (*19*):

1. Gas phase transport of reactants to the substrate
2. Adsorption of reactants onto the surface
3. Photodecomposition of reactants
4. Diffusion of photoproducts and reactants on the surface
5. Nucleation
6. Desorption of products
7. Transport of products away from the surface

Of course, these processes can be similar to those occurring in conventional reactors (*160*), but the presence of photons further increases the complexity of the kinetics. In most photo-CVD reactors, a gas or vapor is adjacent to the surface and the optical source will fragment a portion of the precursors as they migrate to the surface; then, if the surface itself is irradiated, further decomposition will occur in the adlayer (step 3). As we shall see later, evidence has accumulated that indicates that photons can have a profound influence on processes 2, 4, and 6 as well.

As an example of the influence of an optical field over the individual mechanisms in the growth process, let us first consider adsorption and nucleation. One practical ramification of adlayer photolysis is *prenucleation* (*161*), or *photonucleation* (*162*), a process in which laser photodeposition from a chemisorbed adlayer controls the nucleation of metal films. That is, surfaces that have small sticking coefficients for gas phase metal atoms, for example, can be "seeded" by photolyzing metal-alkyl adlayers with a focused UV laser beam. The resulting metal atoms on the surface act as nucleation sites on which further deposition can occur. This process, which overcomes nucleation barriers that often exist at surfaces, has been exploited to deposit patterned metal films with excellent spatial resolution (161–163) (< 4-μm feature widths). In growing crystalline ZnO films by laser CVD, Shimizu et al. (164) showed that the interaction of 193-nm radiation with an NO_2 adlayer on a sapphire substrate *prior to the actual film growth run* was essential for obtaining the highest quality epitaxial films. This effect was attributed to nucleation and its role in the early stages of film growth. Other beneficial effects of irradiating the substrate with VUV photons during film growth include an increase in the growth rate, apparently partially due to photoinduced carriers at the surface, and improved crystalline structure, which presumably arises from optically enhanced surface diffusion.

Once a precursor is adsorbed onto a surface, disposing of the ligands before the element of interest is incorporated into the growing film is of primary concern. The incomplete removal of CO groups from metal hexa-carbonyls (165) has been suggested as the cause for the contamination of Cr films by carbon and oxygen. Houle (165) demonstrated that irradiating the surface during the growth of Cr films from $Cr(CO)_6$ with a 257-nm laser beam (frequency-doubled Ar^+ laser) accelerates the loss of CO from the surface. A number of other experiments have suggested that laser radiation stimulates the desorption of radicals from a surface (166). For both TMA (167) and DMCd (156), 193-nm photons efficiently remove hydrocarbon radicals such as CH_3 from surfaces but 248-nm radiation does not. Since obtaining films of high purity is contingent upon removing the precursor's ligands before they are trapped in a deposit, it is not surprising that the composition of photodeposited films is also usually dependent upon the wavelength of the optical source. As was alluded to in Section 2.1, a hybrid approach to obtaining high purity metal films has proven to be successful. In this case, photochemical and photothermal processes are combined, usually for the purpose of obtaining patterned deposits. Photolysis (photo-dissociation) serves to initiate deposition (and spatially defines the deposit), and pyrolysis carries the process forward. This two-step approach is parti-cularly valuable in those instances where photodissociation is not capable of entirely removing the ligands from the precursor, but the combination of thermal and optical processes results in metal films whose resistivities are close to the bulk value.

The effect of 5-eV (248-nm) and 6.4-eV (193-nm) photons on $Cd(CH_3)_2$ chemisorbed onto oxidized Si is portrayed in Figure 29 (156). Despite the higher optical fluence used at 248 nm, no effect on the chemisorbed layer is observed for the lower energy photons. The rapid decrease in the spectral intensity that is recorded for 193-nm irradiation of the surface is attributed

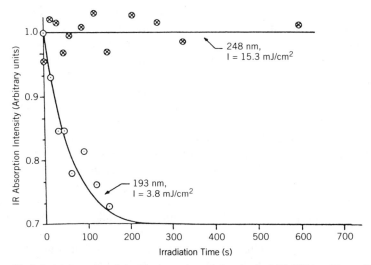

Figure 29. Relative intensity of the IR absorption of chemisorbed DMCd (see Figure 28) as a function of time when the adlayer is irradiated with either 248- or 193-nm photons (reprinted from ref. *156*).

to the removal of hydrocarbon radicals from the surface. Desorption of CH_3 and C_2H_5 radicals also appears to significantly impact the photo-CVD growth of CdTe films (*129*).

The interaction between surface processes such as desorption and adsorption can be represented by a rate equation analysis. Hanabusa and co-workers (*168*) grew Al films on Si by photodissociating dimethylaluminum hydride (DMAlH) with a deuterium lamp or ArF laser and qualitatively confirmed the dependence of film growth rate on intensity by expressing the time rate of change of DMAlH molecules on the surface as (*168*):

$$\frac{dN}{dt} = Rs - \frac{N}{\tau} - A_p N - A_d N. \tag{32}$$

where N is the DMAlH number density; R is the rate at which the precursor molecules impinge on the surface; s is the sticking coefficient; τ is the average residence time of a DMAlH molecule on the surface; and A_p and A_d are the surface photodissociation probability and optically induced surface desorption rates, respectively. Clearly, two of the terms are dependent upon the optical intensity and, by comparing their growth rate data to Eq. (32), Hanabusa et al. (*168*) estimated τ to be 0.22 s. To summarize, measurements of the photo-CVD growth of an extensive number of materials suggests that the presence of optical radiation at the surface, in addition to fragmenting the precursor, promotes surface diffusion and the desorption of products from the surface. The interplay between all of these processes is dependent upon system parameters such as source intensity and gas flow rates and partial pressures.

Not only does the adlayer–surface interaction manifest itself in (a) a broadened absorption spectrum and (b) increased absorption strengths, but it also alters the photochemistry of the precursor relative to that for the gas phase (157, 159, 166, 167, 169–171). Zhang and Stuke (170) investigated the photochemistry of metal-alkyl adlayers by laser time-of-flight mass spectrometry and showed that photodissociation of chemisorbed adlayers of TMA (on Si or quartz) at 248 nm yields much higher abundances of $AlCH_3$ radicals than is measured for the gas phase process. Furthermore, photodissociation of adsorbed aluminum-bearing alkyls at 308 nm occurs quite readily, in contrast with the gas phase species, which is transparent at that wavelength (170). This is consistent with studies that found the cross section for the photodissociation of OCS at 222 nm on surfaces at low coverage to be 3–4 orders of magnitude larger than the gas phase value (172). Braichotte and van den Bergh (169) have also pointed out the advantageous nature of adlayers for the efficient multiphoton dissociation of polyatomics. Since the diffusion rates for photolytic intermediates [such as $Cr(CO)_4$] are lower in an adlayer than in the gas phase, such species in the adlayer are more likely to absorb a second photon from the optical source before diffusing out of the optical beam. Consequently, fragmentation of the precursor is likely to be more complete than in the gas phase.

In leaving this chapter, it must be remembered that, as noted previously, the physics and chemistry underlying all photodeposition processes are complex and, in many cases, are presently poorly understood. The processes occurring at the surface are particularly involved, and the interaction between individual mechanisms depends in a subtle manner on all of the key system parameters (source intensity and wavelength, identity of the precursor, surface composition, etc.). Combining laser spectroscopic and surface analytical techniques has clarified some aspects of surface kinetics in the presence of light, but this area will clearly be the subject of intensive scrutiny for years to come.

CHAPTER

3

REACTORS, OPTICAL SOURCES, AND
ASSOCIATED EQUIPMENT

All photo-CVD systems include these elements: (1) an optical source (lamp or laser) and beam delivery scheme; and (2) a reactor consisting of (a) gas flow equipment for delivering the desired precursor(s) to the substrate at a specified pressure and mass flow rate, (b) an exhaust system, and (c) a substrate and, in most cases, a heated susceptor. Aside from the first item, photo-CVD reactors are quite similar to conventional MOCVD or CBE systems, with the prominent exception of provisions made for introducing photons into the reactor. Consequently, the remainder of this chapter will concentrate on the characteristics of commercially available optical sources and the trade-offs associated with different approaches for delivering the optical beam to the gas or adlayer immediately adjacent to the substrate surface.

3.1. REACTOR CHARACTERISTICS

Even though all reactors have several common components, Figures 30 and 32–37 illustrate the diversity of experimental *configurations* that have been employed thus far for photo-CVD. A simplified schematic diagram of a reactor for the photo-CVD growth of CdTe (*173*) is given in Figure 30. The precursors (diethyltellurium and dimethylcadmium, in this case) are entrained in hydrogen and introduced upstream of the deposition region. The substrate is mounted on a graphite susceptor heated by a quartz halogen lamp, and the unreacted gases are exhausted at the rear of the reactor. A 1000-W Hg–Xe arc lamp provides the necessary UV radiation for photodissociating the alkyls, but Ahlgren et al. (*173*) have introduced a clever modification to the system. By reflecting the arc lamp beam off of a dichroic mirror, radiation outside the 200–250 nm region is rejected and only photons in the ~ 5.0–6.2 eV range reach the surface. This provision minimizes unnecessary heating of the substrate by the optical source.

In order to introduce the UV or VUV radiation from the source into the reactor, an entrance window must be provided, and the material required is dependent upon the wavelengths involved. Since the reactor of Figure 30

47

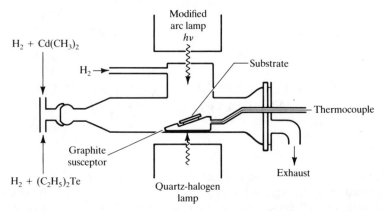

Figure 30. Schematic diagram of a photo-CVD reactor for the growth of CdTe and HgTe epitaxial films [reprinted by permission from Ahlgren et al. (*173*)]. In this arrangement, the optical source was a 1000-W Hg–Xe arc lamp whose spectrum was modified by reflecting the lamp radiation off of a dichroic mirror that was highly reflecting in the 200–250 nm region.

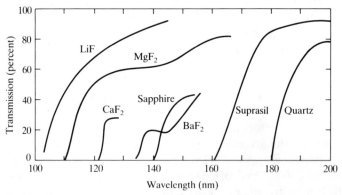

Figure 31. Transmission of readily available window materials in the VUV: LiF, 1.95 mm thick; MgF_2, 1 mm thick; CaF_2, 0.25 mm thick; sapphire, 0.32 mm thick; BaF_2, 1.95 mm thick; Suprasil (enhanced UV transmission quartz), 10 mm thick; and quartz, 2 mm thick. (Reprinted by permission of The American Institute of Physics, John Wiley & Sons, Heraeus–Amersil, Inc., and Corning. Copyright 1965 by The American Institute of Physics.)

utilizes wavelengths in the 200–250 nm region, the optical beam can be propagated efficiently through air and quartz is an acceptable window material. For wavelengths below ~ 180–190 nm, air absorbs strongly (owing to the Schumann–Runge bands of O_2) and different crystalline materials are required for reactor windows. The optical transmission curves for readily available window materials in the VUV are shown in Figure 31. Most are fluoride crystals, and the material offering maximum transmission at the shortest wavelengths, lithium fluoride, "cuts off" above 1000 Å (50% transmission at

~ 1120 Å). One of the materials in Figure 31 is generally more than adequate for most photochemical deposition processes, and one situation requiring the sub-180-nm VUV radiation generated by a deuterium (D_2) lamp is illustrated schematically in Figure 32 (*174*). Calcium fluoride transmits the D_2 lamp radiation that produces a beam having an intensity (for wavelengths below 180 nm) of 12 mW·cm^{-2}. Note that not only are new materials required in this spectral range for transmitting and focusing the incoming radiation, but it is necessary that the entire optical path be evacuated or (in this case) purged with N_2 gas. The latter is only effective for wavelengths down to 150–160 nm. Below ~ 1050 Å, no windows are available and it is necessary to resort to a windowless configuration such as that shown in Figure 33 (*174, 175*). Although it allows one to utilize short-wavelength optical sources, a potential disadvantage of this approach is that the plasma generating the VUV radiation is not isolated from the thin film precursor (Si_2H_6 in this case). An expanded view of the VUV light source based on a disk-shaped plasma that was developed by Yu and Collins (*176*) is shown schematically in Figure 33b. When the light source is operated with H_2, the Lyman alpha transition at

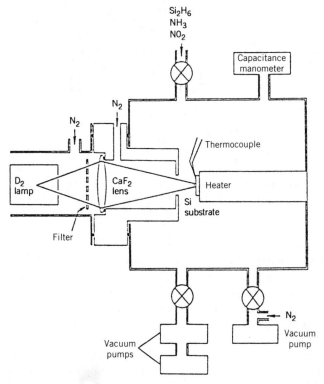

Figure 32. Photo-CVD apparatus of Watanabe and Hanabusa for depositing silicon oxynitride films with a deuterium lamp [reprinted with permission from Watanabe and Hanabusa (*174*)].

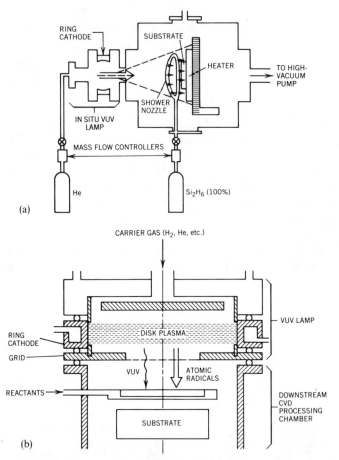

Figure 33. (a) Diagram of the two-chamber windowless He lamp apparatus developed by Zarnani et al. (*175*) (reprinted by permission); (b) detailed view of the VUV light source (H, He, or N) developed for large-area deposition by Yu and Collins (*176*).

121.6 nm dominates the emission spectrum and primarily He^+ (He II) lines are observed when helium is the carrier gas. Strong H_2 molecular emission at ~ 160 nm can also be generated with a lamp of this type.

The lamp parameters and other system operating conditions that were used (*176*) in depositing AlN and amorphous Si films are given in Table 2. Several electrical, optical, and structural characteristics of AlN films deposited by the VUV lamp or an ArF laser (193 nm) are compared in Table 3 (*176*). The properties of the films deposited with the aid of the lamp are generally comparable to or exceed those for films deposited by other techniques.

Table 2. Operating Conditions for the Deposition of Aluminum Nitride or Amorphous Hydrogenated Silicon (a-Si:H) with a windowless VUV Lamp

Films	AlN	a-Si:H
Light source	H_2 lamp H: 121.6 nm	He lamp He^+: 121.5 nm
Cathode current	0.2 A	0.5 A
Cathode voltage	500 V	600 V
Reactant	TMA/NH_3	Si_2H_6
Gas flow	1/40 NH_3: 60 sccm H_2: 200 sccm	Si_2H_6: 20 sccm He: 200 sccm
Total pressure	1 Torr	1–1.5 Torr
Substrate temperature	100–400°C	50–400°C
Deposition rate	60–200 Å·min^{-1}	> 200 Å·min^{-1}

Source: Reprinted from Yu and Collins (*176*), by permission.

Table 3. Properties of AlN Films Photodeposited with a Windowless VUV Lamp or ArF Laser as the Optical Source

	VUV Lamp–CVD	193-nm Laser–CVD
Refractive index	1.7–2.0 (100–400°C)	1.7–1.9 (200–400°C)
Dielectric constant (at 1 MHz)	7.5–8.0	7.0–8.0
Breakdown voltage ($MV·cm^{-1}$)	2–4 (200–400°C)	2–3 (300–400°C)
Resistivity (Ω·cm)	5×10^{13} (400°C)	10^{13} (400°C)
Wet etch rate (Å·s^{-1} with H_3PO_4)	460–500 (100–400°C)	500–1000 (200–400°C)

Source: Adapted from Yu and Collins (*176*).

A detailed diagram of a commercially available low-pressure photo-CVD reactor (Tystar PVD 1000) designed for the deposition of SiO_2, Si_3N_4, and silicon oxynitride (SiO_xN_y) dielectric films is given in Figure 34. Based on the decomposition of mixtures of SiH_4 and N_2O or NH_3 by mercury photo-sensitization, this reactor includes two growth chambers capable of independently processing nine 3-in. wafers or five 4-in. wafers. The optical source is an array of low-pressure Hg lamps or a single grid lamp, and the substrate is heated by four quartz lamps. Depending on the grade of quartz chosen for

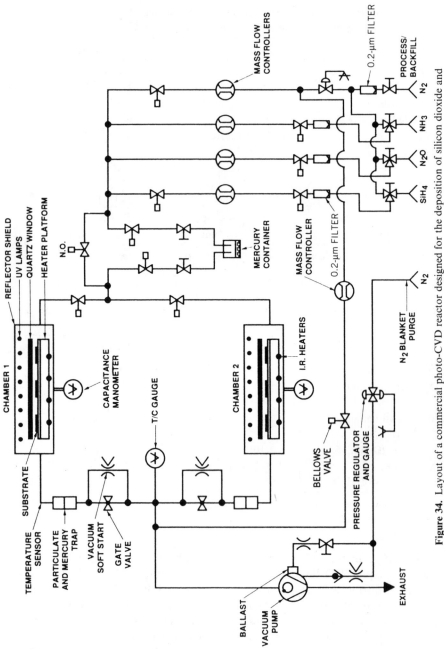

Figure 34. Layout of a commercial photo-CVD reactor designed for the deposition of silicon dioxide and silicon nitride dielectric films by Hg photosensitization (reprinted by permission of Tystar Corp., licensee of Hughes Aircraft Co.—1989).

52

Table 4. General Characteristics of a Commercially Available Photo-CVD reactor (Tystar PVD 1000) Based on Hg Photosensitization

1. Recommended gas flow rates		
SiO_2	Silane (SiH_4)	2 sccm
	Nitrous oxide	60 sccm
	Nitrogen (pump injection)	42 sccm
	Total gas flow rate	104 sccm
	Reactor chamber pressure	1 Torr
	Substrate temperature	150°C
Si_3N_4	Silane	2 sccm
	Ammonia (NH_3)	50 sccm
	Nitrogen (pump injection)	42 sccm
	Total gas flow rate	94 sccm
	Reactor chamber pressure	1 Torr
	Substrate temperature:	150°C
2. Mercury	Triple distilled, reagent grade	500 g
3. Substrate heater	Infrared quartz lamps $4 \times 375\,W = 1500\,W$ total Heat-up time to 150°C: 10 min	

Source: After Schuegraf (*177*).

Table 5. Typical Process Parameters for the Deposition of SiO_2 and Si_3N_4 Films by Hg Photosensitization in a Commercial Reactor

	Film	
Parameter	Silicon Dioxide	Silicon Nitride
Thickness uniformity:		
Within wafer	$\pm 3\%$	$\pm 10\%$
Across heater substrate	$\pm 10\%$	$\pm 15\%$
Repeatability	$\leqslant \pm 3\%$	$\leqslant \pm 5\%$
Deposition temperature	50–200°C	100–200°C
Deposition pressure	0.3–1.0 Torr (40–133 Pa)	0.3–1.0 Torr (40–133 Pa)
Cycle time (150 nm)	45 min	55 min.
Maximum deposition rate	$15\,nm \cdot min^{-1}$	$6.5\,nm \cdot min^{-1}$

Source: After Schuegraf (*177*).

the window, transmission at 254 nm (resonance line of Hg) ranges from ~ 50 to 90%. Tables 4 and 5 give details of the reactor's operating characteristics and film growth process parameters for both SiO_2 and Si_3N_4 films (177). For a reactor chamber pressure of ~ 1 Torr, the maximum SiO_2 deposition rate is 2.5 Å/s. Although Hg is incorporated into the resulting films, concentrations are in the parts per billion (ppb) range.

Broad area deposition of dielectric films with a *laser* is generally accomplished with a parallel configuration such as that illustrated in Figure 35. By avoiding direct irradiation of the substrate, transient heating of the growing film by the laser is eliminated. Because of the ~ 1-cm^2 cross-sectional area and large divergence of most commercially available excimer laser beams, it is generally necessary to compress the rectangular beam in one dimension and collimate it with a simple optical telescope prior to directing the radiation into the reactor. Although the apparatus of Figure 31 was developed by Solanki and Collins (178) for the deposition of dielectric films, Zinck et al. (179) have applied a similar scheme to the growth of epitaxial CdTe films on GaAs. Note also in Figure 35 that provision is made for a low-intensity beam, directed onto the substrate, to stimulate surface reactions (discussed in Chapter 2, Section 2.4).

Section 2.2.2 of Chapter 2 discussed the increased probability for photoionizing a precursor molecule as the optical source wavelength is reduced (that is, $h\nu$ is increased), and particularly as one moves into the VUV. That is, at higher optical frequencies, the relative yield of atomic or molecular ions increases with respect to neutral products. One can exploit the presence of

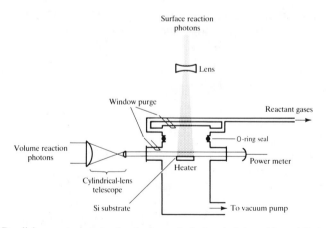

Figure 35. Parallel geometry reactor for the laser photochemical deposition of ZnO films [reprinted from Solanki and Collins (178), with permission].

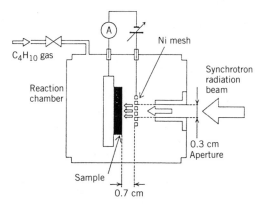

Figure 36. Reaction chamber developed by Ohashi and co-workers for the photo-CVD deposition of films with synchrotron radiation. Positive ions produced near the substrate are accelerated in a weak (< 200 V·cm^{-1}) electric field [reprinted by permission from Ohashi et al. (62)].

these ions to improve the film growth rate or morphology by introducing a weak electric field to the region immediately above the substrate. Figure 36 illustrates the apparatus described by Ohashi and co-workers (62) to grow carbon films with VUV radiation from a synchrotron. Ions produced by the irradiation of n-C_4H_{10} are attracted to the substrate surface, resulting in the growth of hydrogenated carbon films. For a fixed radiation dosage ($hv \geqslant 30$ eV; 500 mA·h of integrated ring exposure), the film growth rate rose linearly with increasing electric field and, for a field strength of ~ 170 V·cm^{-1}, the film thickness was more than an order of magnitude greater than the zero field value. A similar approach in parallel geometry was reported by Geohegan and Eden (68) to deposit thin films of aluminum, thallium, or indium by dissociatively photoionizing metal-halide precursors to generate metal-halogen ion pairs (i.e., M^+–X^-, where M and X represent metal and halogen atoms, respectively). In their apparatus (see Figure 37), the substrates were arranged on a negatively biased plate and the metal cations produced in the gas phase above the substrate were drawn to the substrates under the influence of the electric field.

Regardless of the specifics of reactor design, one problem to be dealt with in all photochemical vapor deposition processes is the tendency for films to grow on the optical window in addition to the substrate. A flow of "purge" gas (typically H_2) across the interior surface of the window will minimize or eliminate this problem (see Figures 30, 32, and 35). An oil coating on the window or *in situ* etching of the window during photodeposition with XeF_2 or discharge-generated F atoms have also been demonstrated successfully (180, 181).

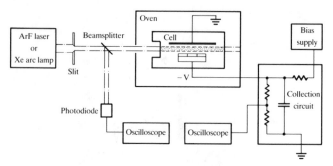

Figure 37. Schematic diagram of the apparatus employed by Geohegan and Eden (68) to deposit thin thallium, aluminum, or indium films by the photoproduction of metal (M) and halogen (X) ion pairs (M^+-X^-).

3.2. OPTICAL BEAM DELIVERY

The optimal manner for delivering the optical beam to the reactor is dependent upon the desired application. Figure 38 qualitatively illustrates the four most common beam delivery configurations (28). The first two (shown as Figure 38a,b) are intended for situations requiring spatially selective deposition. The simplest approach, the *spot focus* or (more commonly known as) the *direct write configuration*, entails scanning a focused laser beam across the substrate. Deposition processes based on this approach generally involve photochemical interactions with the adlayer. If we assume the laser beam to be Gaussian and in the fundamental mode (TEM_{00}), then the diffraction-limited spot size is given by

$$W \simeq \frac{2\lambda}{\pi} \frac{f}{d} \approx 0.64\,\lambda\,\frac{f}{d} \tag{33}$$

where λ is the laser wavelength, and f and d are the focal length and diameter of the focusing lens, respectively. Given a high-quality UV laser beam such as the second harmonic of the Ar^+ laser 514-nm line (257 nm), for example, feature sizes below 1 μm can readily be deposited. The frequent drawback of direct writing is its low scanning speed (typically $\ll$ 100 μm-s^{-1}) (81), although Nambu et al. (182) have recently written $\sim$ 3-μm-wide tungsten lines on Si substrates at speeds up to 300 μm/s. Patterned deposition can be obtained by projection, an approach in which a mask is optically imaged onto the substrate. Figure 39 gives an example of a photo-CVD projection apparatus comprising a reactor and optical system capable of depositing patterned aluminum films with 4-μm resolution (163).

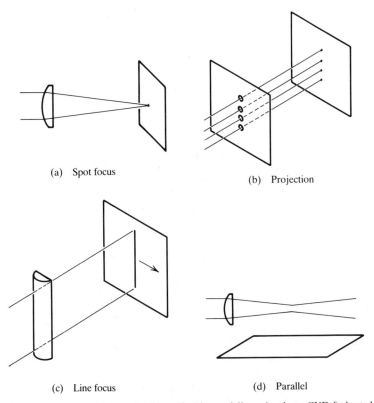

(a) Spot focus

(b) Projection

(c) Line focus

(d) Parallel

Figure 38. Four most common modes for optical beam delivery in photo-CVD [adapted with permission from Houle and Allen (28)]. The first two are designed for the deposition of small features, whereas c and d are intended for large-area deposition applications.

Techniques c and d in Figure 38, the line focus and parallel geometries, respectively, are designed for large-area deposition. For the line focus the optical beam (or substrate) is scanned, but for the parallel configuration the optical beam does not irradiate the surface. The primary advantage of the latter is that the laser intensity can be increased at will without the possibility of thermally damaging the substrate. Of course, it is necessary that the distance from the beam to the substrate be consistent with the lifetime (radiative and collisional) of the transient species produced in the optical beam. Since the collisional mean free path at a pressure of 1.0 Torr is on the order of a millimeter, the photofragments generated by the laser must be sufficiently stable so as to reach the substrate.

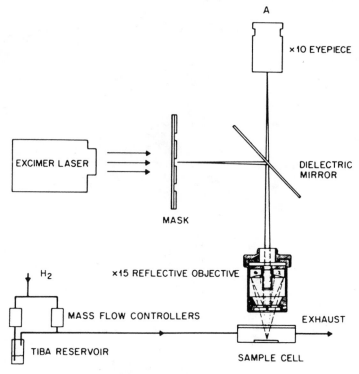

Figure 39. Schematic diagram of a photo-CVD system for the patterned growth of Al films by photodissociating adlayers of triisobutylaluminum (TIBA) with an excimer laser [reprinted by permission from Higashi and Fleming (*163*)].

3.3.   OPTICAL SOURCES

A variety of continuous (CW) and pulsed lamps and lasers is available for photochemical vapor deposition. Several of the most widely used sources— their output wavelengths and power capabilities—will be discussed in this section.

3.3.1.   Lamps

The relatively low cost, high duty cycle, and reliability of discharge lamps have already made them attractive for commercial applications. Low intensity is the only major drawback of lamps for broad-area deposition, which relegates them to driving single-photon absorption processes. Since the poor beam

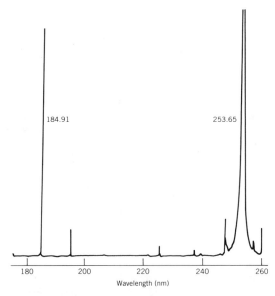

184.91   253.65

180   200   220   240   260

Wavelength (nm)

Figure 40. UV emission spectrum of a low-pressure Hg discharge lamp.

quality of lamps is also an issue for applications involving spatially delineated deposition, this area has thus far been the sole domain of lasers.

Figure 40 shows the UV emission spectrum of a low-pressure mercury discharge lamp. The spectrum is dominated by two lines, the most intense lying at 253.7 nm and the other in the VUV at 184.9 nm. As discussed earlier, the 253.7-nm radiation is effective in triggering the deposition of elemental or compound films by photosensitization processes. Because of their utility in commercial applications other than photo-CVD (such as curing and steriliza-tion), these lamps are commercially available in both linear and grid con-figurations with output powers (at 254 nm) as high as 8 W (*183*).

The spectra emitted by several other commercially available UV lamps are given in Figure 41. Broad continua, extending from the infrared to wavelengths below 250 nm, are produced by xenon and high-pressure Hg lamps. The quartz halogen lamp also generates a blackbody spectrum, but its low color temperature results in weak output below 350 nm, the region where the majority of precursors absorb (see Table 1). Also shown in the figure is the deuterium (D_2) lamp spectrum, which *peaks* for $\lambda \simeq 180$ nm and produces negligible radiation in the visible. The "crossover" wavelength at which the D_2 lamp output exceeds that of the 150-W Xe lamp is ~ 200 nm.

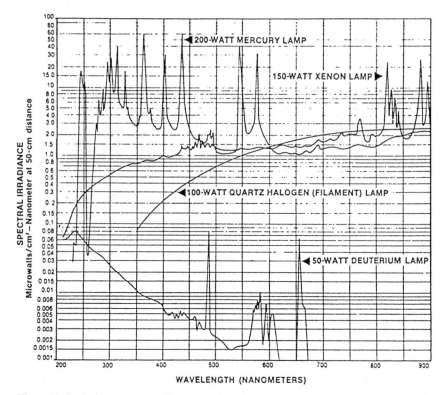

Figure 41. Typical output spectra for several UV-emitting lamps that are available commercially: deuterium (D₂), quartz halogen, xenon, and high-pressure mercury lamps (reprinted by permission of the Oriel Corporation).

Shorter wavelength (VUV) lamps include hydrogen and the rare gases (Figures 42 and 43, respectively) (*69, 184, 185*). Hydrogen emits a highly structured spectrum consisting of two bands (Lyman and Werner) peaking near 120 and 160 nm. In contrast, microwave discharges in the rare gases Ar, Kr, and Xe at high pressure (~ 200 Torr) emit continua in the VUV that arise from electronic transitions of the transient molecules Ar_2, Kr_2, and Xe_2. Notice from Figure 43 that the spectral width of each of the continua exceeds 200 nm.

A recent but important entry into the VUV lamp family is the microwave-excited rare gas–halide lamp. Kumagai and Obara (*186*) have generated 29 W of average power at 193 nm from ArF molecules produced in 2% F_2, 2% Ar, 48% He, and 48% Ne mixture discharges. The average deposited microwave power was 655 W, for an intrinsic efficiency exceeding 4%. Similar results have

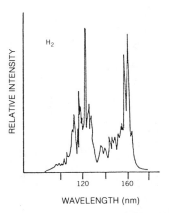

Figure 42. Hydrogen discharge lamp emission spectrum in the VUV [reprinted from Brehm and Siegert (*184*), by permission of Springer-Verlag].

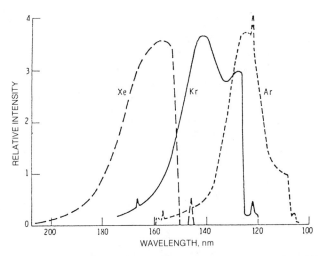

Figure 43. Continua emitted in the VUV by microwave-excited discharges in ~ 2 Torr of argon, krypton, or xenon. The bulk of each continuum is produced by short-lived rare gas dimer molecules [reprinted from Calloway et al. (*69*) by permission].

also been obtained by the same group at 248 nm with emission from KrF. Because of their efficiency and potential for operating at several discrete wavelengths in the UV and VUV, these newly developed lamps will undoubtedly play an increasing role in future photo-assisted deposition processes.

Table 6 summarizes the key properties of several of these lamps. Owing to

Table 6. Summary of the Properties of Several Lamps and Lasers of Importance to Photochemical Vapor Deposition

Source	λ (nm)[a]	CW or Pulsed	Output Power (W)[b]	Comments
a. Lamps				
He ion	121.5	CW		
Hydrogen (H$_2$)	120, 160	CW	10^{-3}–10^{-2}	Discharge excited; H$_2$ pressure $\simeq 2$ Torr; wavelengths cited are the peaks of highly structured bands arising from electronic transitions of the molecule.
Deuterium (D$_2$)	170–270	CW	$\leqslant 60$	Continuum peaking in 180–200 nm region.
Rare gas (Ar, Kr, Xe)	115–130 (Ar) 125–150 (Kr) 147, 150–175 (Xe)	CW	$< 10^{-3}$	Microwave excited; 200 Torr of rare gas.
Mercury (high pressure)	270–400	CW	$\leqslant 1000$	Hg atomic line spectrum (240–600 nm) superimposed onto a continuum; irradiance in 200–250 nm region at 50 cm from a 1-kW lamp is 1100 μW·cm^{-2}; 2100 μW·cm^{-2} in 250–300 nm interval.
Mercury (low pressure)	184.9, 253.7	CW	Linear (germicidal) lamps: up to 8 W at 254 nm, 80 μW·cm^{-2} intensity at 1 m; 7500 h lifetime. Grid configuration: typically 12–15 mW·cm^{-2} (254 nm) at 2.5 cm; up to 25 × 40 cm^2 area	Efficient but low-power lamps emitting primarily at the two indicated wavelengths—generally used in photosensitized reactions involving Hg vapor but also useful in direct photodissociation experiments.

	Wavelength (nm)	Mode	Power	Remarks
Xenon arc	300–900	CW	≤1000	Continuum extending from ~190 to 750 nm having a color temperature of 6000 K; irradiance at 50 cm from 1-kW lamp in 200–250 nm region is 80 μW·cm^{-2}.
Rare gas–halide excimer	193	Pulsed	20–30	Microwave excited; intrinsic efficiency >4% for 655 W of deposited power.
b. *Lasers*				
F$_2$	158	Pulsed	0.8	Up to 3 MW of *peak* power has been reported by Yamada et al. (187) from a research device; for commercial lasers, 1.4 W (average power) is obtainable at 100 Hz.
Rare gas–halides				
ArF	193	Pulsed	15	25 W is generated at a pulse repetition frequency (PRF) of 100 Hz; 55 W at 300 Hz.
KrCl	222	Pulsed	6	12 W at 100 Hz.
KrF	248	Pulsed	28	100 W is available at 300 Hz.
XeCl	308	Pulsed	16	70 W at 300 Hz; longest laser gas mixture lifetimes of any of the rare gas–halides.
Ar$^+$	333.6–363.8	CW	7.0	Low efficiency but excellent beam quality suitable for direct write applications; several atomic lines are lasing in the indicated spectral region.

Table 6 (*Continued*)

Source	λ (nm)[a]	CW or Pulsed	Output Power (W)[b]	Comments
Ar^+	275.4–305.5	CW	1.5	Recent improvements in wall materials in the laser have permitted higher discharge currents and correspondingly larger output powers in this region than were available previously; in a research device, 20 mW of output power at 231.5 nm was also reported (188).
Ar^+	257	CW	$\lesssim 0.2$	Intracavity doubling of the 514.5 nm line of the Ar^+ laser with a newly available nonlinear crystal, β-BaB_2O_4, yields 1–2% conversion efficiencies for 514.5 nm powers of 8–10 W.
Copper vapor (510.6, 578.2 nm) —summed in β-BaB_2O_4	271	Pulsed	0.2–0.5	As indicated, the copper vapor laser emits on two lines at 511 and 578 nm: commercial models produce up to 100 W of average power (both lines combined) at pulse repetition frequencies up to 10 kHz; β-BaB_2O_4 enables one to efficiently sum these two wavelengths to produce 271 nm.[c]

[a]Wavelengths cited are those at which peak emission occurs or, in the case of continua, the wavelength region that lies roughly between the 50% points.
[b]Average values that are commercially available are quoted; for the excimer laser, the pulse repetition frequency is taken to be 50 Hz.
[c]This laser is not widely used at the present time in photo-CVD but is likely to grow in importance owing to its high duty cycle. In addition to generating almost 0.5 W at 271 nm by summing the green and yellow Cu vapor laser lines, recent experiments have produced > 0.6 W at 255.1 nm by frequency doubling the laser's 510.6 nm line.

64

their low intensities, UV and VUV lamps are most useful for those processes in which reliability, duty cycle, and low cost, rather than peak power, are the driving considerations.

3.3.2. Lasers

Tremendous progress in the engineering development of VUV and UV lasers has been made over the past decade. Despite the fact that the energies required for the rupture of chemical bonds within a polyatomic molecule ($\sim$ 3–6 eV) generally dictate the need for UV photons, prior to 1975 few efficient lasers existed in this spectral region. However, the discovery and commercial development of the rare gas–halide lasers has dramatically transformed that situation by making available photons ranging in energy from 3.5 to 6.4 eV ($193 \leqslant \lambda \leqslant 351$ nm) in beams having peak intensities (unfocused) exceeding 10 MW·cm^{-2}. In fact, much of the early work in photo-CVD was driven by the availability of these laser sources. It is not surprising, then, that a large fraction of the photochemical processes to be discussed in Chapters 4–7 rely on an excimer laser as the optical source. Figure 44 indicates the spectral breadth of the wavelengths produced by the rare gas–halide and diatomic halogen lasers (such as the F_2 laser). In most instances, the power available from commercial discharge devices is given.

The current specifications for commercially available F_2 (158 nm), ArF (193 nm), KrF (248 nm), and XeCl (308 nm) lasers are also summarized in Table 6 (187). Over the last few years, the major improvements that have been made in excimer laser performance, such as gas fill lifetime and laser pulse repetition frequencey (PRF) have permitted the introduction of these lasers into the industrial environment. The progress made in the reliability of the rare gas–halide lasers is reflected in Table 7, which lists the current lifetimes (in millions of pulses at 200–300 mJ/pulse) for critical components of the laser.

Another workhorse of photo-assisted deposition has been the argon ion (Ar$^+$) laser. Although generally operated on the visible lines lying between 454.5 and 528.7 nm (with the strongest transition at 514.5 nm), this laser also produces significant ouput power on several discrete lines in the UV (275–364 nm). Recent advances in the materials used in contact with the high current discharge have permitted the power deposited into the plasma to be increased, with a resulting improvement in output power (188). At present, commercial models are available that will generate 1.5 and 7.0 W of CW power on two separate groups of lines in the 275.4–305.5 nm and 333.6–363.8 nm regions, respectively. Further improvements in plasma tube and resonator materials will certainly lead to increases in the output powers given in Table 6. In addition, a number of photo-CVD experiments and direct write studies,

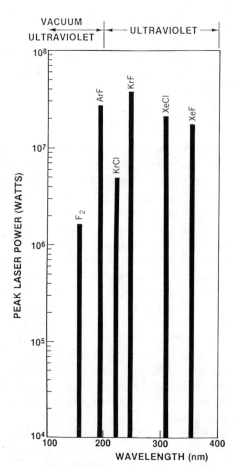

Figure 44. Wavelengths of several of the rare gas halide lasers in the UV. The fluorine molecular laser in the VUV (158 nm) is also shown and the peak powers given for each laser are those available from commercial systems.

Table 7. Lifetimes of Various Components of Commercially Available Excimer Lasers

Component	Laser transition	Lifetime (10^6 pulses)
Gas	KrF (248 nm)	70
	XeCl (308 nm)	100
Optics (cleaning)	KrF	100
	XeCl	100
Discharge chamber	KrF	> 500
	XeCl	> 500
High-voltage switch (thyratron)	KrF	> 1000
	XeCl	> 1000

in particular, have employed the second harmonic (intracavity doubled) of the 514.5-nm line at 257 nm.

These UV lasers are well suited for those photodeposition applications in which the source intensity and beam quality are key considerations. Despite their relatively high cost and (for pulsed lasers) low duty cycle operation, these characteristics, as well as the capability to quickly change the output wavelength, make them attractive in photo-assisted processes for which lamps are not suitable.

The recent introduction of a new nonlinear crystal (beta barium borate, β-BaB$_2$O$_4$) has converted one other well-known visible laser into an intense UV source suitable for photo-CVD. The pulsed copper vapor laser emits on two visible lines (511 and 578 nm) that would normally be of little interest for most photodeposition processes. However, summing both lines in β-BaB$_2$O$_4$ yields up to ~ 0.5 W at 271 nm ($\hbar\omega = 4.6$ eV). Although little has yet been done with this new source, the high repetition frequencies (> 10 kHz) and average powers available from this laser make it attractive for future work.

3.3.3. Other Sources

Although conventional lamps and pulsed lasers are most widely used for photo-CVD, several new optical sources have been reported. Windowless lamps, based on high-current, low-voltage (~ 0.5 A, 200–600 V) dc discharges in gases such as He and N$_2$, have been shown to efficiently generate VUV and deep-UV radiation (H, 121.6 nm; He; 121.5 nm; N$_2$, 120–400 nm) (*146, 175, 176, 189*) and were discussed in Section 3.1 in connection with Figure 33. Also, the increasing availability of synchrotrons, coupled with their ability to produce high-brightness, tunable UV and VUV radiation, has prompted several demonstrations of synchrotron radiation-assisted deposition of elemental (*62*) and compound (*190*) films.

CHAPTER

4

METAL FILMS

Because of their varied applications to microelectronics, metal films have been at the center of much of the effort in photochemical vapor deposition. This chapter will review the elements that have been deposited to date, with emphasis given to the necessary growth parameters. For each group of metals, the various approaches that have been taken to photodepositing films will be briefly described, and the current status of film growth rates and purity are summarized in Table 8 (191–234). The latter is not intended to be exhaustive but rather to reflect the range of deposition conditions and results that have been demonstrated to date for a particular film.

4.1. GROUP IIB METALS (Cd, Zn)

Photodeposition of the Group IIB metals cadmium and zinc from their respective alkyls has been the subject of intense scrutiny (the photodeposition of mercury will be discussed in Chapter 5 in conjunction with the II–VI compound semiconductor films). In 1966, Jones et al. (200) demonstrated that thin cadmium films (< 100 Å thick) could be patterned on SiO_2 substrates by photodissociating dimethylcadmium (DMCd) with 254-nm radiation from an Hg resonance lamp. Subsequent work has centered on the photodissociation of DMCd in the gas phase or in an adlayer with excimer or Ar^+ (257 nm; second harmonic of 514.5 nm) radiation. In a series of papers, Deutsch, Ehrlich, and Osgood (27, 161, 191, 201) reported the broad area deposition of Cd films as well as the writing of lines with widths as small as 0.7 μm. DMCd pressures of 1–4 Torr in the reactor allowed deposition rates of 7–13 Å·s^{-1} to be obtained by photodissociating chemisorbed and physisorbed aklyl molecules on glass and quartz substrates. At higher intensities, deposition rates up to 1000 Å·s^{-1} have been achieved. Table 8 summarizes these results as well as the experiments carried out by Rytz-Froidevaux et al. (16, 158) with a Kr^+ ion laser to deposit Cd films on quartz.

Similar studies have been carried out for zinc where dimethylzinc [$Zn(CH_3)_2$] or diethylzinc [$Zn(C_2H_5)_2$] acted as the gas phase precursor. Johnson and Schlie (225) successfully deposited several square centimeters of

Table 8. Summary of Metal Films Deposited by Photochemical Vapor Processes

Element	Precursor	Optical Source	λ(nm)	Experimental Conditions	Refs.	Comments
Al	Al(CH$_3$)$_3$	Ar$^+$ (SH)^a	257	3 Torr TMA, 750 Torr He; 1–60 s exposure times; ~2·10^{-4} W of UV laser power	27,161, 191,192	Deposition and prenucleation demonstrated with TMA.
Al	Al(CH$_3$)$_3$	ArF	193	50 Torr TMA and 22.5 Torr GeH$_4$ were photodissociated to yield heavily Al-doped Ge films; laser intensity = 4.5 MW·cm^{-2} ($\perp$ geometry); laser PRF^b = 20 Hz; He buffer pressure $\simeq$ 380 Torr; quartz substrates	193	Al doping levels up to 1 atom % were obtained; experiments carried out at room temperature.
Al	Al(CH$_3$)$_3$	ArF	193	$\perp$ geometry	27	> 100 Å·s^{-1} growth rates.
Al	Al(CH$_3$)$_3$	KrF	248	$\parallel$ geometry; laser beam 1 mm above a polysilicon substrate; TMA/H$_2$ gas mixtures, 1:3 ratio, 100 sccm mass flow rate; total pressure = 0.5 Torr; substrate temperature = 200°C	194	Deposition rates up to 1000 Å·min^{-1} and film thicknesses over 5000 Å were obtained (see Table 9). Metal step coverage was found to be conformal.
Al	Al(CH$_3$)$_3$	KrF	248	TMA pressure: 10^{-3}–1 Torr; laser energy: 20–80 mJ/pulse, 10 Hz; laser focused to ~100 μm dia. spot at edge of Si substrate; growth time: 6·10^3–4·10^4 laser pulses (10 min–1 h)	195	Substrate was inclined with respect to laser beam to prevent adlayer photolysis; growth rates near laser beam exceeded 500 Å·min^{-1} for 0.5 Torr TMA; typical deposition rate $\simeq$ 100 Å·min^{-1}
Al	Al(CH$_3$)$_3$	Microwave-excited rare gas lamps	100–200	Quartz, Si, and sapphire substrates at room temperature; rare gas lamps at 300 Torr and driven by 100 W, 2.45 GHz microwave generator; lamp fluence: 5·10^{15}–5·10^{16} cm^{-2}·s^{-1}, 0.2–10.0 Torr TMA vapor	69	450 ± 50 Å thick Al films were deposited on Si with Xe or Kr lamps and TMA mass flow rate and pressure of 100 sccm and 0.2 Torr, respectively. Increased deposition was observed upon biasing substrate negative by 200 V. Deposition was attributed to photoproduction of (CH$_3$)$_3$Al$^+$ and (CH$_3$)$_2$Al$^+$ ions.

	Precursor	Wavelength (nm)	Source	Conditions	Ref.	Comments
Al	Al(i-C$_4$H$_9$)$_3$	257	Ar$^+$(SHJ)[a]	1–5 mW of 257 nm laser power; ~4 μm spot size at substrate; counterpropagating CO$_2$ laser for pyrolytic deposition	162	Demonstrated catalytic activity of photodeposited Al films.
Al	Al(i-C$_4$H$_9$)$_3$	248	KrF	~1 Torr TIBA in H$_2$ carrier gas; substrate temperature $\simeq$ 250°C; laser fluence $\simeq$ 20 mJ·cm^{-2} ($\perp$ geometry)	163	Projection patterning of Al films on Si, SiO$_2$, Al$_2$O$_3$, and GaAs demonstrated. Growth rates from 30 to 1000 Å·min^{-1} were obtained.
Al	Al(i-C$_4$H$_9$)$_3$	193	ArF	$\perp$ geometry; laser focused to 2 mm spot on Si(100) substrate; laser energy density $\simeq$ 10 mJ·cm^{-2}; PRF = 40 Hz	196	Photonucleation of MOCVD growth of Al demonstrated.
Al	AlH(CH$_3$)$_2$	257	Ar$^+$(SHJ)[a]	60–100 mW of laser power; 1–2 Torr dimethylaluminum hydride, focal dia. of beam ~2.3 μm; substrate: 0.3 μm SiO$_2$ on Si	81	For scanning speeds of 1 μm·s^{-1}, Al film deposition rates of 0.09 μm·s^{-1} were obtained for ~115 kW·cm^{-2} laser intensity; lines written at 10–50 μm·s^{-1} were thin (200–500 Å thick) with resistivities 2–10 times larger than for bulk Al.
Al	AlH(CH$_3$)$_2$	193, 160–240	ArF laser, D$_2$ lamp (150 W)	VUV intensity (λ < 180 nm): 140 mW·cm^{-2}; $\leqslant$ 15 mJ/pulse at 193 nm, 10 Hz; substrate temp: 100–250°C; $\perp$ geometry; Si, quartz substrates.	168	Deposition rates (200°C): 190 Å·min^{-1} (lamp), 380 Å·min^{-1} (laser, 230 mW·cm^{-2}); VUV illumination allows films to be deposited at temperatures lower than required for pyrolysis; deposition rate saturates at ~0.6 Å/pulse for ArF pulse energies $\gtrsim$ 8 mJ; electrical resistivity (270°C): 6.2 $\mu\Omega$·cm (VUV), 140 $\mu\Omega$·cm (thermally deposited film).
Al	AlH(CH$_3$)$_2$	160–240	150 W D$_2$ lamp	n-Type Si(100) or quartz substrates; 130–270°C substrate temperature; lamp intensity ($\lambda \leqslant$ 180 nm): 140 mW·cm^{-2}; DMAH vapor pressure at 25°C: 2 Torr; DMAH pressure in reactor: 5·10^{-5} Torr	197	At 200°C, deposition rate = 200 Å·min^{-1}; Film resistivities down to 6.2 $\mu\Omega$·cm (2.3 × bulk value) were obtained; thermally deposited films very resistive (140 $\mu\Omega$·cm); maximum deposition rates exceeded 300 Å·min^{-1} (200°C).

71

Table 8 (*Continued*)

Element	Precursor	Optical Source	λ (nm)	Experimental Conditions	Refs.	Comments
Al	AlI$_3$	ArF laser	193	Laser intensity $\simeq 5$ MW·cm^{-2}; PRF = 30 Hz; Ni substrates	68	Thin (several hundred Å) films grown by photoionizing AlI$_3$ in an electric field.
Au	Dimethylgold(III) acetylacetonate [Me$_2$Au(acac)]	Hg lamp, XeCl, KrF, ArF	254, 308, 248, 193	Typical laser pulse energies: 24 mJ: 1,000–20,000 pulses; 4X optical projection system; 0.25 cm × 0.25 cm patterned image at quartz substrate	198	2000 Å thick Au films were deposited with line widths of 2 μm; highest gold contents observed at 248 nm (KrF); carbon content > 50% at 193 and 308 nm; no deposition at 351 nm (XeF), showing process to be photochemical.
Au	(CH$_3$)$_3$Au:P(CH$_3$)$_3$	KrF	248		199	Broad area and patterned films were deposited.
Cd	Cd(CH$_3$)$_2$	Hg arc lamp (200 W)	254	Quartz substrate at room temperature; 10 Torr dimethyl-cadmium; 10 s exposure time	200	Patterned deposition was demonstrated by projecting onto the substrate through a metal mask; film thickness was estimated to be < 100 Å.
Cd	Cd(CH$_3$)$_2$	ArF laser	193	Single-pulse energy = 1–10 mJ (30–300 mJ·cm^{-2}), 10 ns pulses; ~ 1 Torr DMCd; average power = 20 mW·cm^{-2}	27, 161, 191, 201	Broad-area deposition ($\perp$ geometry) on glass and SiO$_2$ substrates; beam was apertured by ~ 200 μm dia. pinhole; addition of He to DMCd reduces deposition rate.
Cd	Cd(CH$_3$)$_2$	Ar$^+$ laser (*SH*)^a	257	3 Torr DMCd buffered with 750 Torr He; 0.1–3.0 mW at 257 nm (0.08–10^4 W·cm^{-2})	27, 161, 191, 192, 201–205	Direct writing of Cd lines on glass and SiO$_2$; line widths as low as 0.7 μm were obtained. Deposition rates: ~ 7 Å·s^{-1} for 1 W·cm^{-2} and 4 Torr DMCd; up to 1000 Å·s^{-1} at 10^4 W·cm^{-2}.
Cd	Cd(CH$_3$)$_2$	Kr$^+$ laser	337–676	DMCd adsorbed onto quartz substrates	16, 157	Demonstrated broadening of DMCd adlayer absorption into the red.

		Laser	Wavelength (nm)	Conditions	Ref.	Comments
Cr	$Cr(CO)_6$	F_2, ArF, KrF, XeCl	158, 193, 248, 308	⊥ geometry; deposition over 2.5 cm × 2.5 cm area at room temperature or 150°C; quartz, Pyrex, or Si substrates	206, 207	Best results were obtained at 248 nm. Film properties are summarized in Table 11. Adherence properties of the films were improved by heating the substrate to 150°C prior to exposing films to air. Substrate precleaning with oxygen atoms produced by dissociating O_2 with excimer laser prior to deposition improved adhesion.
Cr	$Cr(CO)_6$	KrF	248	Room temperature; laser intensities $\leq 7\ \mathrm{MW \cdot cm^{-2}}$	208	Deposited films are smooth with ~50% Cr; remainder C and O.
Cr	$Cr(CO)_6$	ArF, KrF Ar⁺ (SH)[a]	193, 248, 257	Si, quartz, and Pyrex substrates; room temperature	98	Deposition of lines and metal dots was demonstrated.
Cr	$Cr(CO)_6$	Frequency-doubled, tunable dye laser	280–350	0.02–0.20 mJ/pulse laser energies; Si and quartz substrates; room temperature, ⊥ geometry	209	Increase in deposition rate for $\lambda <$ 350 nm observed because of transition between one- and two-photon processes; deposition rate $>$ 100 Å·min⁻¹ at $\lambda = 284$ nm and 0.1 mJ/pulse laser energy.
Cr	$Cr(CO)_6$	Ar⁺ (SH)[a]	257	1–2 mW of 257 nm power focused to ~8 or 160 μm dia. spot; exposure times: 10–15 min; 0.1–0.2 Torr of carbonyl precursor; Si and sapphire substrates	210, 211	Photochemistry of adsorbed and gas phase carbonyl compared; all CO ligands are not removed from adsorbed molecule; composition of films determined.
Cr	$Cr(CO)_6$	ArF, KrF, XeCl	193, 248, 308	8–12 mJ/pulse, 10–60 Hz; parallel and ⊥ geometries; room temperature; Pyrex, quartz, Al substrates	94	Deposition rate 300 Å·min⁻¹ at 10 Hz (308 nm); rate independent of laser repetition frequency: 0.05, 0.28, and 0.10 Å/pulse at 308, 248, and 193 nm, respectively. Good film adherence only in ⊥ geometry.

73

Table 8 (*Continued*)

Element	Precursor	Optical Source	λ (nm)	Experimental Conditions	Refs.	Comments
Cu	Bis-(1,1,1,5,5,5-hexafluoropentanedionate)-copper(II) [Cu(Hfac)$_2$]	Hg lamp, Ar$^+$ (SH),[a] ArF, KrF lasers	254, 257, 193, 248	Quartz and Si substrates; $\perp$ geometry; laser intensities: Ar$^+$(SH): 0.2–2.0·10^4 W·cm^{-2} (10 μm dia. focal spot size); ArF, KrF: 1–10 MW·cm^{-2} (100 Hz, 50 s exposure times); Hg lamp: 5 mW·cm^{-2} (254 nm)	212	Film deposition with Hg lamp observed only with Cu(Hfac)$_2$/ethanol vapor mixtures; deposition rates: with Ar$^+$ (257 nm) — <1–200 Å·min^{-1}, $\sim$1 μm thick films, $\sim$10–90% carbon; with ArF or KrF lasers, no deposition below 1 MW·cm^{-2}; film thicknesses limited to $\sim$500 Å at higher intensities.
Cu	Cu(Hfac)$_2$	Hg lamp, Ar$^+$ (SH),[a] KrF	254, 257, 248	Laser intensities: Ar$^+$(SH), ≤10^4 W·cm^{-2}; KrF, ≥1 MW·cm^{-2}; Hg lamp, 5 mW·cm^{-2} at 254 nm; exposure times: Hg lamp, few hours; Ar$^+$(SH), 10–40 min; KrF, 1 min at 100 Hz; quartz, Si/SiO$_2$ and carbon on mica substrates	213	Deposition rates obtained were a few Å·s^{-1} for Ar$^+$ (SH) source; exposures were adjusted to obtain 500 Å–1 μm thicknesses; ≤5% carbon concentration was obtained for Ar$^+$ (SH) illumination (see Table 12).
Fe	Fe(CO)$_5$	Hg–Xe arc lamp (2.5 kW)		Glass and silverfoil substrates at room temperature; Fe(CO)$_5$ pressure: 10^{-2} Torr; $\perp$ geometry; irradiation period: 10^4 s	214	Thin conducting Fe films were deposited; maximum film thickness was $\sim$300 Å; deposition was attributed to photoelectrons ejected from substrate by the optical source.
Fe	Fe(CO)$_5$	ArF, KrF	193, 248	$\perp$ geometry; substrate at 77 K or room temperature; laser peak intensity: 5·10^5–10^7 W·cm^{-2}; repetition frequency ≤100 Hz; Fe(CO)$_5$ pressure: $\sim$10^{-7}–$\sim$10^{-3} Torr; GaAs(100) and sapphire substrates	215	High-quality Fe films were obtained at a laser intensity of $\sim$5·10^6 W·cm^{-2} at 100 Hz repetition frequency; it was found to be necessary to photolyze carbonyls at a rate that is faster than their rate of arrival at the substrate. Atomic C and O concentrations were found to be $\sim$7% and 4%, respectively.

74

Element	Compound	Laser/Source	Wavelength (nm)	Conditions	Ref.	Comments
Fe	$Fe(CO)_5$	$Ar^+ (SH)$,[a] ArF, KrF	254, 193, 248	Pyrex, quartz, and Si substrates at room temperature; $Fe(CO)_5$ pressure: ~ 0.5 Torr; laser intensity: up to $600\ W\cdot cm^{-2}$	98	Fe lines were deposited at rates up to $30\ \text{Å}\cdot s^{-1}$ for $600\ W\cdot cm^{-2}$ of laser intensity.
Fe	$Fe(CO)_5$	Hg lamp	254	GaAs(100) or Si substrates; substrate temperature: 77 K; $Fe(CO)_5$ pressure: $\leqslant 33$ Torr	216, 217	Photodeposited films contain C and O, which are removed by heating film to room temperature.
Fe/Ni	Ferrocene, nickelocene	ArF, N_2 lasers	193, 337	Laser pulse energies and repetition rates: N_2, 4 mJ, 50 Hz ($0.3\ J\cdot cm^{-2}$); ArF, ~ 100 mJ ($0.1\ J\cdot cm^{-2}$), 10 Hz; quartz substrate; ferrocene and nickelocene powder in reactor	218	Deposits were $\sim 40\ \mu m$ wide, 7000 Å thick for N_2 laser; 6 mm wide, 4000 Å thick for ArF laser; film composition: 92% Fe, 8% Ni (193 nm), 65% Fe, 35% Ni (337 nm).
Ga	$Ga(CH_3)_3$	Xe arc lamp (300 W)		$20\ W\cdot cm^{-2}$ of UV intensity over 0.25–$0.50\ cm^2$ area	219	Deposition on GaAs substrates yielded Ga droplets with carbon impurity.
Ga	$Ga(CH_3)_3$	$Ar^+ (SH)$[a]	257	Frequency-doubled, CW, or mode-locked Ar^+ laser; TMG pressure: 58 Torr; $\perp$ geometry; CW intensities: 120–400 $W\cdot cm^{-2}$; mode-locking intensities: 0.15–1.75 $kW\cdot cm^{-2}$	100	Film deposition rates of $17\ \text{Å}\cdot s^{-1}$ at $230\ W\cdot cm^{-2}$ of laser intensity were measured for quartz substrates. Deposition rates up to $70\ \text{Å}\cdot s^{-1}$ were observed.
In	$In(CH_3)_3$, InC_5H_5	Xe arc lamp		Room temperature	219	Metal droplets deposited on Pyrex—smoother films on GaAs; films deposited on Pyrex from the cyclopentadienyl contained substantial carbon.
In	InI	ArF	193	InI number density: $\simeq 10^{15}\ cm^{-3}$; ArF intensity: $\simeq 5\ MW\cdot cm^{-2}$	68	Thin In films were deposited on nickel-plated stainless steel by dissociatively photoionizing InI. Film thicknesses of several hundred Å were obtained at deposition rates of $\sim 420\ \text{Å}\cdot h^{-1}$.

Table 8 (*Continued*)

Element	Precursor	Optical Source	λ (nm)	Experimental Conditions	Refs.	Comments
Ir	Ir(acac)$_3$	Nd:YAG (frequency quadrupled)	266	Si(111) substrate (room temperature); laser beam 0.5 cm from substrate; laser energy: 5 mJ/pulse; peak intensity: 32 MW·cm^{-2} (20 ns pulses, 1 mm focal dia.); ~ 1.1 Torr of Ir(acac)$_3$	39	Multiphoton ionization of precursor yields predominantly Ir$^+$ ions and metallic stripe films.
Mn	Methylcyclopentadienyl manganese	Hg–Xe arc lamp (2.5 kW)		Glass or silverfoil substrates; vapor pressures up to 10^{-2} Torr; $\perp$ geometry	214	Deposited Mn films nonconductive; film deposition attributed to photoelectrons produced at substrate by lamp.
Mo	Mo(CO)$_6$	F$_2$, ArF, KrF, XeCl	158, 193, 248, 308	Deposition at room temperature or 150°C; $\perp$ geometry; He purge on window: quartz, Pyrex, or Si substrates	206, 207	Film properties ($\lambda = 248$ nm) summarized in Table 11.
Mo	Mo(CO)$_6$	ArF, KrF	193, 248	8–12 mJ/pulse; 10–60 Hz; $\perp$ geometry	94	Deposition rates: 0.15 and 0.13 Å/pulse at 193 and 248 nm, respectively.
Mo	Mo(CO)$_6$	Ar$^+$ laser	350–360	3 μm laser beam spot size; 15–65 mW laser power; Mo(CO)$_6$ pressure: 0.17 Torr; glass and GaAs substrates at 90°C; writing speeds (typical): 1.0–1.8 μm·s^{-1}	220	Deposition rates up to 2700 Å·s^{-1} were observed for 30 mW of laser power, 1 μm·s^{-1} writing speed, and a glass substrate; film thicknesses up to 0.8 μm were obtained with resistivities ~ 100 to 5000 times bulk value; for GaAs substrates, film thicknesses were < 0.2 μm.
Pb	Pb(C$_2$H$_5$)$_4$	Ar$^+$ (*SH*)^a	257	Quartz substrate at room temperature; tetraethyl lead pressure: 0.5–3.8 Torr; laser intensity: ~ 160–640 W·cm^{-2}	116	Deposition rate is linear in both laser intensity and Pb(C$_2$H$_5$)$_4$ pressure and deposition rates up to 30 Å·s^{-1} were observed for a pressure of 0.7 Torr.

Element	Precursor	Light source	Wavelength	Conditions	Ref.	Comments
Pb	Pb(C$_2$H$_5$)$_4$	Hg arc lamp (200 W); Hg low-pressure lamp (6 W)		Tetraethyl lead pressure: typically 0.2 Torr; irradiation time: 39–70 min	221	Thin lead films were deposited on a quartz crystal microbalance; deposition rates were small: ~2 monolayers/min.
Pb	Pb(C$_2$H$_5$)$_4$	Hg arc lamp (250 W); Hg low-pressure lamp (15 W)		Glass substrate at 0°C; tetraethyl-lead pressure: ~5·10^{-2} Torr	222	Pb films were deposited by photo-dissociating adsorbed alkyl precursor molecules. The deposition rate on already deposited Pb was 3 times larger than for Pb(C$_2$H$_5$)$_4$ adsorbed on "clean" glass; Pb(C$_2$H$_5$)$_4$ adsorbed on lead was shown to absorb radiation at longer wavelengths (280–290 nm) than would the gas phase species.
Pt	Pt(acac)$_2$	Nd:YAG (frequency quadrupled)	266	Si(111) substrate at room temperature; Pt(acac)$_2$ partial pressure: 1.1 Torr; 5–10 mJ laser energy/pulse; 550 μm beam focal diameter; peak intensity: 105 MW·cm^{-2} (20 ns pulses).	39	Metal stripes, ~330 μm in width, were deposited on Si; composition: ≥95 atom % Pt, <5 atom % C, O, and H; primary species produced was determined by time-of-flight mass spectroscopy to be atomic Pt.
Pt	Pt(PF$_3$)$_4$	KrF	248	⊥ geometry; quartz and graphite substrates; Pt(PF$_3$)$_4$ pressure: ≈0.6 mbar (0.46 Torr).	223	Film deposition on quartz requires a laser fluence ≥50 mJ·cm^{-2}; films are composed of microcrystallites.
Pt	Pt(CH(CO·CF$_3$)$_2$)$_2$, platinumbishexa-fluoroacetylaceto-nate [Pt(Hfacac)$_2$]	Ar$^+$	350–360	Glass and GaAs substrates; substrate temperature: 110°C; laser power: ≤65 mW; laser scan velocity: 0.4 μm·s^{-1}; precursor pressure: 0.3 Torr.	220	Pt metal lines up to ~500 Å thick were deposited on glass and GaAs; film resistivities were 2–10 times the bulk value for glass substrates but >10 times the bulk value when GaAs is the substrate.
Pt	Pt(Hfacac)$_2$	Ar$^+$ (SH)^a	257	Quartz substrate	118	

Table 8 *(Continued)*

Element	Precursor	Optical Source	λ (nm)	Experimental Conditions	Refs.	Comments
Pt	Pt(Hfacac)$_2$	Ar$^+$	350–360	Intensity at substrate: 10^{-2}–10^3 kW·cm^{-2}; glass substrates at room temperature; metal organic vapor pressure: typically $2.3 \cdot 10^{-3}$ Torr; laser beam dia. at substrate: 2.5 μm	169, 224	Film deposition in the ~ 10^{-2} to 10 kW·cm^{-2} intensity range is photolytically driven. Deposition rates rise linearly to a maximum value of ~ 120 Å·min^{-1} at 10 kW·cm^{-2}. Deposition is attributed to adlayer photolysis. Above 50 kW·cm^{-2}, photolytic deposition is rapidly followed by pyrolytic growth as the thin photodeposited metal film is heated by the laser. At high writing speeds (> 5 μm·s^{-1}) and ≲ 200 mW of laser power, ~ 5 μm wide Pt metal stripes up to ~ 0.1 μm in height are deposited photolytically.
Pt	η^5-C$_5$H$_5$Pt(CH$_3$)$_3$	XeCl, Ar$^+$	308, 351, and 364	Precursor vapor pressure: ≃ 0.33 Torr; glass, quartz, sapphire, and GaAs substrates; laser energy: 2.6 mJ/pulse; 10 Hz repetition frequency; 10 min irradiation period; ⊥ geometry; CW (Ar$^+$) laser: 4.5 mW·mm^{-2} for 10 min	117	Film deposition rate of 100 Å·min^{-1} was obtained; carbon concentration ~ 3.5 atom %; ~ 100 μm wide lines were deposited, and no deposition occurred outside of irradiated region.
Se	Se(CH$_3$)$_2$	Hg/Xe arc lamp (1000 W)	~ 220–300	1–10 Torr Se(CH$_3$)$_2$, 0–620 Torr Ar	225	Yellowish-red Se films were deposited at rates up to 1 Å·s^{-1} over ~ 2 cm dia. Total thicknesses up to ~ 0.6 μm were obtained.
Sn	Sn(CH$_3$)$_4$	Ar$^+$ (SH),^{a} Hg lamp	257.254	257 nm power: ~ $2 \cdot 10^{-4}$ W; 1–60 s exposure time	191	Few details given regarding deposited films.

78

Element	Precursor	Source	Wavelength (nm)	Conditions	Ref.	Comments
Sn	$Sn(CH_3)_4$	Ar^+(SH)[a]; Hg lamp	257, 254	Quartz and carbon membrane substrates; 257.3 nm laser intensity: $3.8 \cdot 10^2$–$2.6 \cdot 10^3$ $W \cdot cm^{-2}$; exposure time: 15–30 s; 10–60 Torr $Sn(CH_3)_4$ vapor pressure; Hg lamp power: 3 mW (254 nm)	118	For low 257 nm intensities ($\lesssim 10^3$ $W \cdot cm^{-2}$), tin films are amorphous, but at $\sim 2.5 \cdot 10^3$ $W \cdot cm^{-2}$ polycrystalline films are deposited; the deposition rate is linear in precursor pressure. The film deposition rate measured with Hg lamp excitation is $6 \cdot 10^{-2}$ $Å \cdot min^{-1}$; the spatial resolution achievable by this process appears to be at least 0.2 μm.
Sn	$Sn(CH_3)_4$ $Sn(C_4H_9)_4$ $SnBr_2(C_4H_9)_2$ $SnCl_4$	Hg lamp array	254	GaAs(001) substrates were irradiated at normal incidence by the lamp in presence of tin-containing precursor for ~ 30 min	226	Thin (< 30 Å) Sn films were deposited to study the Sn/GaAs interface.
Ti	$TiCl_4$	Ar^+(SH)[a]	257	5 mW power (laser is mode-locked at 41 MHz; laser beam spot size: 4 μm; intensities: 50 $kW \cdot cm^{-2}$; $TiCl_4$ pressure: < 10 Torr	227	Ti lines of ~ 4 μm width written on $LiNbO_3$; waveguides fabricated; deposition rate = 900 $Å \cdot s^{-1}$ for laser intensity of 40 $kW \cdot cm^{-2}$.
Tl	TlI	Xe arc lamp (150 W)	~ 220–300	Silver substrate; lamp intensity at substrate: $\simeq 0.5$ $W \cdot cm^{-2}$	68	Photoionization of TlI yielded thin Tl films: deposition rate ~ 400 $Å \cdot h^{-1}$. Pronounced diffusion of Tl into underlying silver.
W	$W(CO)_6$	Ar^+	350–360	3 μm laser beam spot size; laser power: 15–65 mW; glass, GaAs, and sapphire substrates; substrate temperature: 90°C; 1 $\mu m \cdot s^{-1}$ scanning speed; $W(CO)_6$ pressure: 0.035 Torr	220	Film thicknesses up to ~ 1100 Å were obtained on glass with 65 mW; for GaAs, thicknesses were limited to < 600 Å; film resistivities were > 200 and $\gtrsim 1.5$ times the bulk value for GaAs and glass substrates, respectively.
W	$W(CO)_6$	F_2, ArF, KrF, XeCl	158, 193, 248, 308	Deposition at room temperature or 150°C; $\perp$ irradiation geometry; He purge on window; quartz, Pyrex, or Si substrates	206, 207	Properties of films deposited at 248 nm are summarized in Table 11.

Table 8 (*Continued*)

Element	Precursor	Optical Source	λ(nm)	Experimental Conditions	Refs.	Comments
W	$W(CO)_6$	ArF, KrF, Ar^+ (SH)[a]	193, 248, 257	Room temperature deposition on Si, Pyrex, and quartz substrates; precursor pressure: 0.2 Torr; laser intensity: $\leqslant 1000\ W\cdot cm^{-2}$	98	Lines with $\sim 2.5\ \mu m$ widths were deposited at rates up to $\sim 3.5\ \text{Å}\cdot s^{-1}$. Scanning speed was $\sim 0.9\ \mu m \cdot s^{-1}$.
W	$W(CO)_6$	Ar^+ (SH)[a]	257	$\leqslant 15\ mW$ of laser power; $\sim 30\ \mu m$ beam diameter; quartz substrate	228	Photodissociation of $W(CO)_6$ shown to be the rate limiting step for photodeposition.
W	WF_6/H_2	ArF	193	Laser power: 4–7 W (average power) at 50 Hz repetition frequency; $\parallel$ geometry–laser beam $\sim 1\ mm$ above Si or Si/SiO_2 substrates; substrate temperature: 200–440°C	229	Deposition rates $> 1000\ \text{Å}\cdot min^{-1}$ were obtained. Resistivity of films grown at 440°C was twice the bulk value.
W	$W(CO)_6$	Xe arc lamp		Deposition on *n*- and *p*-InP and *n*- and *p*-GaAs	199	Ohmic contacts were obtained with photodeposited W films.
W	$W(CO)_6$	ArF, KrF	193, 248	8–12 mJ/pulse; 10–60 Hz; $\perp$ geometry; room temperature and 40°C	94	Deposition rates: 0.03 and 0.01 Å/pulse at 193 and 248 nm, respectively.
W	WF_6/H_2	ArF	193	10 mJ/pulse; $\leqslant 40\ Hz$; quartz and Si substrates	230	Best results at 200°C; 32 Å·min^{-1} deposition rate.
Zn	$Zn(CH_3)_2$	Ar^+ (SH)[a]	257	DMZ on quartz or Pyrex substrates; laser intensity: 0.01–$10^2\ W\cdot cm^{-2}$; film thicknesses up to several micrometers	27,161,204	Writing of $< 1\ \mu m$ width lines; ~ 2–3 μm features deposited by prenucleation.

Metal	Precursor	Source	Wavelength (nm)	Conditions	Ref.	Results
Zn	$Zn(CH_3)_2$	Hg/Xe arc lamp (1000 W)	~220–300	1–10 Torr $Zn(CH_3)_2$, 0–620 Torr Ar buffer; lamp intensity for $\lambda < 245$ nm $= 0.4$ W·cm^{-2}; quartz or sapphire substrates	225	Film thicknesses up to 6000 Å were deposited over several cm^2; wavelength threshold for DMZ dissociation was found to be 245 nm; deposition rate ~1 Å·s^{-1}.
Zn	$Zn(CH_3)_2$	ArF	193	Laser fluence $\simeq 0.9$ J·cm^{-2}; laser apertured by 100–500 μm pinhole	231	Ohmic contacts were formed on p-InP by photodissociating DMZ and heating resulting Zn film with laser.
Zn	$Zn(C_2H_5)_2$	Ar$^+$ $(SH)^a$	257	Glass, quartz, or Si substrates; laser spot size: 3 μm; laser power: 0.15–1 mW; scanning speed: 1.4 μm·s^{-1}; diethylzinc pressure: 1–5 Torr	232	Film thicknesses in excess of 6000 Å were obtained at room temperature on glass. Similar results were observed for Si. Thickness decreased rapidly with increasing substrate temperature, and the deposition rate was linear in laser power up to 0.2 mW at 1–5 Torr of metal-organic vapor.
Zn	$Zn(C_2H_5)_2$	ArF, KrF lasers	193, 248	Excimer laser operated with unstable resonator optics for low (0.5 mrad) beam divergence; laser PRF ≤ 150 Hz, typical value: 22.5 Hz; $\perp$ geometry; energy fluence at substrate: 0.6 J·cm^{-2}; alkyl pressure: several Torr	233	Line (2 μm minimum feature size) and broad-area (~ 20 μm × 20 μm) deposition on InP at room temperature; at 193 nm, deposition due to gas phase reactions–poor spatial resolution. At 248 nm, interaction is primarily with adlayer.
Zn	$Zn(C_2H_5)_2$	Hg lamp	185, 254	$Zn(C_2H_5)_2$ pressure: 10^{-2}–1 Torr; glass substrate at room temperature	234	Deposition rates ranged from 0.1 to 0.6 Å·s^{-1}.

[a] The abbreviation *SH* denotes the second harmonic of the Ar ion laser at 257 nm.
[b] PRF = pulse repetition frequency.

81

Zn films on quartz or sapphire substrates by illuminating $Zn(CH_3)_2/Ar$ gas mixtures with deep UV radiation from a mercury–xenon arc lamp. They showed that the useful portion of the lamp's spectrum lay below 245 nm and obtained film thicknesses up to 0.6 μm. Aylett and Haigh (*233*) deposited Zn lines (widths down to 2 μm) and larger area films (400 μm^2) with an ArF or KrF laser operated in an unstable resonator mode so as to minimize diffraction of the beam. The morphology of the deposits indicated that at 193 nm the primary deposition mechanisms are gas phase processes, whereas at 248 nm ($h\nu = 5$ eV/photon) the photodissociation of alkyl molecules in the adlayer dominates, which leads to deposits of high spatial definition. Film deposition rates as high as 0.6 Å$\cdot$s^{-1} were obtained by Ando et al. (*234*) by photodissociating diethylzinc with a low-pressure mercury lamp (185, 254 nm). As shown in Figure 45, their measurements were made for $Zn(C_2H_5)_2$ pressures ranging from 10^{-2} to 1 Torr.

Limited effort has been directed thus far toward a detailed characterization of photodeposited Cd or Zn films. While several studies exploring the viability of Cd and Zn lines as interconnects have been carried out, both metals have heretofore been principally of interest because of their importance to II–VI compound films such as CdTe and ZnS. Deposition rates observed to date

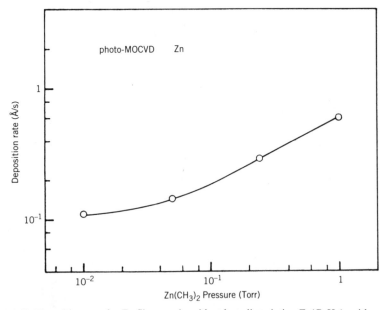

Figure 45. Deposition rate for Zn films produced by photodissociating $Zn(C_2H_5)_2$ with a mercury lamp [after Ando et al. (*234*)].

are typically $1-5 \text{Å·s}^{-1}$, and photodissociation of the alkyl precursor in the gas phase and in the adlayer are clearly both occurring and contributing to film growth. Jackson (235) has shown that monomethylzinc ($ZnCH_3$) is a stable product of the photoexcitation of $Zn(CH_3)_2$ at 248 nm, but is it the critical species in the alkyl fragmentation and film growth process? Larciprete and Borsella (236) have recently clarified the relative merits of dimethylzinc (DMZ) and diethylzinc (DEZ) as precursors by using time of flight (TOF) ion detection techniques. They demonstrated that when DMZ is photodissociated with an ArF or dye laser, one CH_3 ligand is removed but the second frequently is not (as exemplified by the presence of considerable $ZnCH_3$ in the TOF mass spectrum). In the case of DEZ, however, the mass spectrum shows only Zn ions and Zn films deposited by photodissociating DEZ at 248 nm have carbon impurity concentrations below the detectability limit of Auger electron spectroscopy (AES). Films 300–3000 Å thick were deposited at rates up to 30Å·s^{-1} with excimer laser (KrF) fluences as high as $\sim 70 \text{mJ·cm}^{-2}$. Detailed measurements of the electrical and chemical properties of photodeposited Zn and Cd films and in situ spectroscopic studies during film growth will be necessary to determine the quality of photodeposited Zn films and whether resorting to precursors other than the alkyls is desirable.

4.2. GROUP IIIB METALS (Al, Ga, In, Tl)

The Group IIIB metals aluminum, gallium, and indium have also found widespread use in microelectronic and optoelectronic devices as interconnects and as elements in III–V semiconductor compounds. Of the three elements, aluminum has received the most attention in photodeposition studies because of its excellent bulk conductivity. Furthermore, owing to its high vapor pressure [the dimer form of TMA, $Al_2(CH_3)_6$, has a vapor pressure of 8.4 Torr at 20°C] (195), TMA is a relatively convenient precursor to handle. Both direct writing of Al lines and broad area deposition of metal films have been demonstrated. With dimethylaluminum hydride (DMAH) as the precursor and 60–100 mW of 257 nm laser power, for example, Al lines $\sim 2-3 \,\mu m$ in width have been written on a variety of substrates at speeds of $1 \,\mu\text{m·s}^{-1}$ (81). At this scanning speed and a laser intensity of $\sim 115 \text{kW·cm}^{-2}$, deposition rates of $0.09 \,\mu\text{m·s}^{-1}$ were obtained. Broad-area deposition on polysilicon was demonstrated by Solanki et al. (194) by photodissociating TMA in parallel geometry with a KrF laser. Deposition rates up to 1000Å·min^{-1} and film thicknesses $> 0.5 \,\mu m$ were recorded. Perhaps more importantly, the composition of the resulting films was found to be $> 88\%$ Al, $< 5\%$ C, and $< 7\%$ O and the step coverage was conformal. Similar results were reported by Motooka and co-workers (195), who also photodissociated TMA at 248 nm and, upon analyzing their films by AES found no indication of Al—C bonding. These conclusions are consistent with the time-of-flight mass spectrometry experiments of Beuermann and Stuke (41) (discussed in Chapter 2, Section 2.1), which showed that, in photodissociation of TMA, the carbon-bearing $AlCH_3$ radical was not produced beyond the threshold wavelength of 230 nm. Consequently, one expects that films produced by photolyzing TMA at 193 nm will contain considerably more carbon than those deposited with longer wavelength (i.e., 248 nm) radiation. Higashi (237) has succinctly summarized the state of photo-CVD Al films deposited from Al-alkyls by noting that the highest quality films have been produced by combining photolytic and pyrolytic processes. Although photodissociation initiates the growth process by partially decomposing the alkyl and producing nucleation sites, pyrolysis is effective in continuing deposition and in minimizing the incorporation of carbon, in particular. Thus, photolysis of Al-alkyl adlayers is well suited for providing spatial resolution, and subsequent pyrolysis improves film quality and deposition rate.

Aluminum films having resistivities as low as 6.2 $\mu\Omega\cdot$cm (2.3 times the bulk value) were deposited with a deuterium lamp by Hanabusa et al. (197) by photodissociating DMAH with a D_2 lamp. One motivation for exploring this precursor stems from the low carbon concentrations in AlGaAs films grown with DMAH by MOCVD. Hanabusa et al. (197) realized deposition rates up to 200 Å·min^{-1} at 200°C and, in contrast to the photochemically deposited films, Al films deposited by pyrolysis at 260°C were highly resistive (140 $\mu\Omega\cdot$cm). Subsequent work by Hanabusa, Oikawa, and Cai (168) with DMAH compared the results obtained with a D_2 lamp and an excimer laser. Because of the low-absorption cross section of gas phase DMAH at 248 nm ($\sim 7 \times 10^{-20} cm^2$), (81), the laser experiments were carried out at 193 nm (ArF). At 200°C, deposition rates of 190 and 380 Å·min^{-1} were obtained with lamp (140 mW·cm^{-2}) and laser (230 mW·cm^{-2}) radiation, respectively. Not only did the presence of VUV illumination at the substrate permit Al film deposition to occur at considerably lower temperatures than those accessible by pyrolysis, but the VUV once again dramatically improved the film conductivity. Figure 46 displays the resistivity for Al films deposited by laser or D_2 lamp as a function of the substrate temperature (168). Because of the low concentrations of carbon in the photolytically deposited films (< 6%), DMAH is a particularly attractive new precursor for optical source wavelengths below 200 nm. Considering the results obtained by Baum et al. (238) and Sasaoka et al. (239) in thermally decomposing trimethylaminealuminum hydride and diethylaluminum chloride, respectively, to obtain Al lines or patterns, other aluminum hydride or aluminum halide precursors appear to be promising candidates as photo-CVD precursors. Table 9 summarizes the properties of Al films deposited from TMA, TIBA, and DMAH precursors (163, 168, 194).

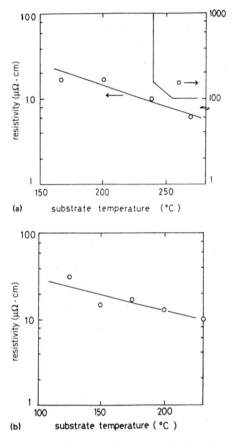

Figure 46. Comparison of the Al film resistivities obtained by Hanabusa et al. (*168*) by photo-dissociating DMAH with either (a) a D_2 lamp or (b) an ArF laser in perpendicular geometry. The lowest observed resistivity (6.2 $\mu\Omega \cdot cm$) is a factor of approximately 2.3 larger than the bulk value. The inset of (a) indicates the resistivity of a thermally grown (CVD) film.

Calloway, Galantowicz, and Fenner (*69*) demonstrated that Al films can also be deposited from TMA with VUV-emitting rare gas lamps. Producing fluences ranging from 5×10^{15} to 5×10^{16} photons/cm$^2 \cdot$s, microwave-excited Xe and Kr lamps generate radiation of sufficiently short wavelength to photoionize the precursor, yielding $(CH_3)_3Al^+$ and $(CH_3)_2Al^+$ ions. Upon biasing the Si substrate negatively by 200 V, increased deposition rates were observed (*69*), and films up to 450 ± 50 Å in thickness were obtained.

Considerably less is known of the photodeposition of Ga and In. Gallium films have been deposited from $Ga(CH_3)_3$ with both arc lamps (*219*) and a

Table 9. Comparison of the Properties of Aluminum Films Photodeposited from Trimethylaluminum (TMA), Triisobutylaluminum (TIBA), and Dimethylaluminum Hydride (DMAH)

	Ref. *194*	Ref. *163*	Ref. *168*	Ref. *168*
Precursor	TMA	TIBA	DMAH	DMAH
Optical Source	KrF Laser (248 nm)	KrF Laser	ArF Laser (193 nm)	D_2 Lamp (140 mW·cm^{-2})
Irradiation geometry	‖	⊥	⊥	⊥
Substrate temperature (°C)	200	250	200	270
Deposition rate (Å min^{-1})	1000	30–1000	80–380	10–290
Resistivity ($\mu\Omega$·cm)	9.1	5	12	6.2
Impurity concentration (%):				
C	< 7	—	2.5	3.3
O	< 5	—	3	3

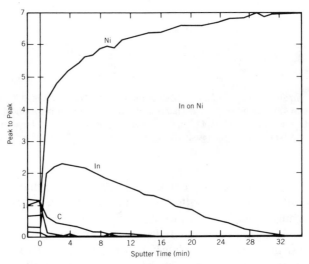

Figure 47. Auger depth profile for an In film deposited onto a nickel substrate at $\sim 300°C$ by dissociative photoionization of InI. Note that the In has diffused more than 0.3 μm into the Ni [after Geohegan and Eden (*68*)].

frequency-doubled Ar^+ laser (100). Deposition rates up to $70\,\text{Å}\cdot s^{-1}$ have been achieved with the latter, whereas the large-area studies of Aylett and Haigh (219) yielded gallium metal droplets. Aylett and Haigh (219) carried out similar experiments with various indium precursors (trimethylindium and cyclopenta-dienylindium) and GaAs and Pyrex substrates. They once again observed the formation of droplets on glass but obtained films with smoother morphology on GaAs. Thin indium films have also been deposited with the apparatus of Figure 37 by dissociatively photoionizing InI at 193 nm to yield $In^+ - I^-$ pairs (68). For an InI number density of $10^{15}\,cm^{-3}$, deposition rates were low, $\sim 400\,\text{Å}\cdot h^{-1}$, but the films were essentially free of iodine because of the selectivity of the photoionization process. An AES depth profile of an In film deposited on a Ni substrate is shown in Figure 47. The diffusion of In into the substrate is pronounced but iodine is conspicuous by its absence. Similar results were obtained for thallium films deposited on silver by dissociatively photoionizing TlI with a Xe arc lamp (68).

4.3. GROUP IVA, VIA, AND VIIA TRANSITION METALS (Cr, Mo, W, Ti, Mn)

Extensive studies of photochemical vapor deposition of Cr, Mo, and W, in particular, have been carried out, with most prior work involving the photo-dissociation of their respective hexacarbonyls with an excimer laser or fre-quency-doubled Ar^+ laser. Large-area ($> 6\,cm^2$) chromium films have been deposited with XeCl (308 nm), KrF (248 nm), and ArF (193 nm) lasers in perpendicular and parallel geometries (94, 206–208), but most efforts have concentrated on the perpendicular configuration because of the poor adhesion of films deposited if the substrate is not irradiated. The highest deposition rates ($2000\,\text{Å}\cdot min^{-1}$) and film purities have been obtained at 248 nm, but the films still contain only 50–60% Cr (208), with the remainder consisting of C and O.

The contamination of metal films photodeposited from a hexacarbonyl precursor is not unique to Cr (210, 240), and Table 10 (240) summarizes the compositions of films resulting from the photodissociation of $Cr(CO)_6$, $Mo(CO)_6$, and $W(CO)_6$ at various wavelengths in the UV (241–243). Detailed studies (210, 240) of the composition and surface morphology of films deposited under varying conditions indicates that partial photodissociation of the hexacarbo-nyl occurs in the gas phase and coordinatively unsaturated metal carbonyls [i.e., $M(CO)_n$, $n < 6$] arrive at the substrate, where further dissociation of the molecule takes place. Both dissociative chemisorption and photodissociation of the metal carbonyls occur on the surface, but in the process the C—O bond is ruptured, apparently by an autocatalytic process (95, 240). For the $Cr(CO)_4$

Table 10. Summary of the Carbon and Oxygen Contaminant Concentrations in Films Deposited by Photodissociating the Hexacarbonyls of Cr, Mo, and W at Various UV Wavelengths

Metal	λ(nm)	Buffer Gas	Composition (%)[a]			Ref.
			Metal	Carbon	Oxygen	
Cr	158–308	He	> 92	0.8	< 7	207
	248	Ar	50	30	20	208
	248	Ar	42(s)	8	50	241
			65(b)	25	10	
	248	Ar	47(s)	27	26	242
	257		31(b)	—	69	228
	257	N_2	33(s,b)	—	66	211
Mo	158–308	He	> 92	0.9	< 7	207
	257		44(b)[b]	6	37	228
	257	N_2	20(s)	40	40	211
			60(b)	15	25	
	Hg lamp	He	60(s,b)	20	20	243
W	158–308	He	> 92	0.7	< 7	207
	257		36(b)	37	27	228
	257	N_2	25(s)	35	40	211
			55(b)	10	35	

Source: Reprinted from Singmaster et al. (*240*), by permission.
[a] Region analyzed is indicated in metal column: (s), surface; (b) bulk following sputtering; —, unknown.
[b] Films also contained 13% nitrogen.

molecule, this sequence is represented in a generalized manner as

$$Cr(CO)_4 + hv \longrightarrow Cr + C + O + 3CO \qquad (34)$$

$$Cr(CO)_4 \longrightarrow Cr + 2C + 2O + 2CO \qquad (35)$$

These reactions account for the fact that Cr(C, O) films produced by Cr(CO)$_6$ photolysis generally have C/O ratios of ~ 1:1, that is, the film stoichiometry is CrCO. As noted by Hess (*95*), the free carbon and oxygen atoms of Eqs. (34) and (35) react with chromium to form Cr_2O_3 and Cr_3C_2 (*95, 240*). The irony of this situation (*95*) is that the hexacarbonyls were initially chosen as precursors for Cr, Mo, and W since it was presumed that, in the presence of UV radiation, the metal–CO bond would preferentially rupture and the volatility of the CO ligand would minimize carbon incorporation into the

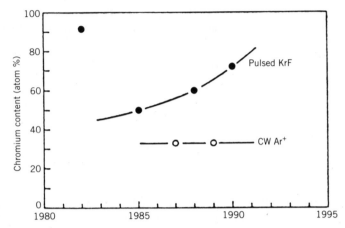

Figure 48. Variation of chromium content in Cr(C, O) films deposited by the photodissociation of $Cr(CO)_6$ at 248 nm (KrF laser) or 257 nm (frequency-doubled CW Ar^+ laser) [reprinted from Hess (*95*), by permission].

metal film. Reality is considerably more involved, but (as shown in Figure 48) steady progress in improving the quality of Cr films deposited by KrF photolysis of $Cr(CO)_6$ has been realized over the past few years. Films having chromium content beyond 70 atom % have recently been produced by adding hydrogen buffer gas to the $Cr(CO)_6$ vapor flow stream (*244*).

Direct writing of Cr lines (and dots) has also been demonstrated (*98*) at 257 nm, and Mayer et al. (*209*) have shown that the increase in deposition rate observed for source wavelengths below 350 nm occurs because of the transition from two-photon to single-photon dissociation of the carbonyl for $hv \gtrsim 3.5\,eV$. Deposition rates of $> 100\,\text{Å}\cdot\text{min}^{-1}$ were obtained for a laser wavelength of 284 nm.

Comments similar to those for Cr films could be made for tungsten and molybdenum as well. Table 11 compares the deposition rates and electrical and structural properties of Mo, W, and Cr films grown with an optical source wavelength of 248 nm. The deposition rates are comparable for all three elements, and the carbon and oxygen impurity levels range from < 10–40% and < 7–40%, respectively. The composition of the Mo-containing films is consistent with an MoC_2O stoichiometry, whereas that for the tungsten films is $WC_{<1}O_{<1}$ (*240*). Also, the proposed chemical kinetics at the surface are quite similar to Eqs. (34) and (35) except that the desorption of CO or CO_2 appears to be quite efficient for Mo deposition.

Tungsten lines ($< 3\,\mu m$ width) and ohmic contacts have been deposited from the carbonyl precursor, and (as indicated in Table 11) broad area

Table 11. Summary of Several Physical and Electrical Properties of Mo, W, and Cr Films Photodeposited with a KrF Laser (248 nm) from Metal Hexacarbonyl Precursors

	Deposition Rate (Å/min)	Resistivity ($\Omega/\square$)	Adherence (PSI)	Tensile stress (dyn/cm^2)
Mo	2500	0.9	> 8000	$< 3 \times 10^9$
W	1700	4.5	> 9400	$< 2 \times 10^9$
Cr	2000	6	> 7800	$< 7 \times 10^9$

Source: After Solanki et al. (207).

deposition rates up to 1700 Å/min were obtained at 248 nm (207). Larger *relative* rates were observed by Flynn et al. (94) at 193 nm. If one resorts to WF_6 (in the presence of H_2) as the precursor, deposition rates in excess of 1000 $\text{Å} \cdot \text{min}^{-1}$ are obtained in *parallel* geometry at 193 nm, with film resistivities within a factor of 2 of the bulk value (229). A frequency-doubled Ar^+ laser, which provided up to 600 W/cm^2 at the substrate, was used by Ehrlich et al. (98) to deposit 2.5-μm-wide metal lines on Pyrex for 900 W/cm^2 of laser intensity and a $W(CO)_6$ pressure of 0.2 Torr. Less work has been reported for the photo-deposition of molybdenum, but the results are comparable to those obtained for Cr and W. Gilgen et al. (220) have deposited molybdenum and tungsten lines on glass, GaAs, and sapphire substrates by scanning the focused UV lines (350–360 nm) from an Ar^+ laser. For $Mo(CO)_6$ and $W(CO)_6$ pressures of 0.17 and 0.035 Torr, respectively, and a writing speed of 1 μm/s, deposition rates up to 2700 $\text{Å} \cdot \text{s}^{-1}$ for Mo lines and W film thicknesses up to 1100 Å were observed. The lowest film resistivities were measured for glass substrates.

Two other transition metals that have been deposited photochemically are titanium and manganese. Direct writing of 4-μm-wide Ti lines on $LiNbO_3$ has been exploited to fabricate single-mode waveguides (227) with only 5 mW of power in the second harmonic of the Ar^+ laser (257.2 nm). Manganese films have also been photodeposited from a methylcyclopentadienyl precursor with a 2.5-kW Hg–Xe lamp (214).

4.4. GROUP IVB METALS (Sn, Pb)

Although tin (118, 191, 226) and lead (116, 221, 222) films have been photo-deposited from alkyl precursors, characterization of the gas and surface pro-cesses and resulting films has been limited. In both cases, adlayer photolysis appears to trigger film deposition and thin (< 100-Å-thick) films have been deposited on GaAs, glass, and sapphire substrates. Chiu et al. (116) have deposited Pb films from tetraethyllead with frequency-doubled Ar^+ laser radiation and obtained deposition rates up to $30 \, Å \cdot s^{-1}$ for a precursor pressure of 0.7 Torr.

Mingxin, Monot, and van den Bergh (*118*) deposited tin films on both quartz and carbon substrates by photodissociating $Sn(CH_3)_4$ with the second harmonic of the Ar^+ 514 nm laser line or an Hg resonance lamp. Depending on the source intensity, both amorphous and polycrystalline films resulted. With Hg resonance lamp (3 mW, 254 nm) dissociation of the metal alkyl, deposition rates were much smaller than those obtained with the laser—typically, $0.06 \, Å \cdot min^{-1}$. Interference structure appearing in the photodeposited films indicates that patterned tin films with features as small as 0.2 μm can be deposited by this approach.

4.5. GROUP VIII TRANSITION METALS (Fe, Pt, Ir)

Conducting films of iron were deposited onto various substrates by George and Beauchamp (*214*) by photodissociating the pentacarbonyl with a Hg–Xe arc lamp. Carried out at $Fe(CO)_5$ partial pressures of 10^{-2} Torr, these studies yielded ~ 300-Å-thick films after an exposure period of 10^4 s. The authors attributed deposition to photoelectrons (ejected from the substrate by the optical source) that decomposed the carbonyl precursor.

Thin iron films have also been deposited under ultra-high-vacuum (UHV) conditions by photolyzing $Fe(CO)_5$ adsorbed onto GaAs or Si substrates. As indicated in Table 8, both excimer lasers and arc lamps have been used in perpendicular geometry, and the films studied to date contain significant concentrations (> 10 atom%) of carbon and oxygen. Iron has also been deposited onto Pyrex, quartz, and GaAs at rates up to $30 \, Å \cdot s^{-1}$ with 257-nm radiation from a frequency-doubled Ar^+ laser (*98*).

In more recent work, Armstrong and co-workers (*218*) deposited Fe/Ni composite films by photodissociating mixtures of ferrocene and nickelocene vapor at room temperature. The ferrocene absorption spectrum peaks near 190 nm, whereas that for nickelocene reaches its maximum value at ~ 285 nm. Upon irradiating the ferrocene/nickelocene mixture at 193 nm (ArF laser), the deposit was found to consist of 92% Fe and 8% Ni. Similar measurements made at 337 nm with a pulsed N_2 laser yielded 65% Fe–35% Ni deposits. Film deposition rates at 193 and 337 nm were determined to be 20 and $> 200 \, Å \cdot s^{-1}$, respectively.

Platinum films have been deposited from a variety of precursors onto quartz, graphite, glass, sapphire, and GaAs substrates. Utilizing both pulsed and CW UV sources, Koplitz et al. (*117*) demonstrated that ~ 1000-Å-thick Pt films can be deposited from $(CpPt(CH_3)_3)$, where $Cp = \eta^5\text{-}C_5H_5$. For 26 mW of average power at 308 nm (10 Hz), films were observed to grow at a rate of $\sim 100\,\text{Å}\cdot\text{min}^{-1}$ over an area of $\sim 3\,\text{mm}^2$. The carbon concentration in these films was found to be $\lesssim 3.5$ atom%, and similar results were obtained for Pt films photodeposited with a CW Ar^+ ion laser operating simultaneously on the 351 and 364 nm lines. The absence of any detectable growth when the alkyl vapor was irradiated with visible Ar^+ laser photons indicates that the deposition process is photochemically initiated. Platinum films have also been deposited on quartz and graphite with KrF excimer radiation and the frequency-doubled 514 nm line of Ar^+ (118, 223, 224).

Gilgen et al. (220) deposited thin Pt lines [from Pt(Hfacac)_2] on glass and GaAs at a scan speed of $0.4\,\mu\text{m}\cdot\text{s}^{-1}$ with the 350–360 nm UV lines of an Ar^+ laser. For glass substrates, the film resistivities were ~ 2–10 times the bulk value. Similar experiments reported by van den Bergh and co-workers (169, 224) delineated the transition from photolytic to pyrolytic deposition of Pt lines. One of several lines available from a CW Ar^+ laser (257, 334, 363, and 454 nm) were used, and experiments were, again, carried out with platinum bishexafluoroacetylacetonate. At low scan speeds, deposition for laser intensities above $50\,\text{kW}\cdot\text{cm}^{-2}$ is primarily pyrolytic in nature. Photolytic deposition, on the other hand, is dominant for a laser power at or below 200 mW and a writing speed of $> 5\,\mu\text{m}\cdot\text{s}^{-1}$; 5-$\mu$m-wide stripes up to $\sim 0.1\,\mu$m in thickness were deposited.

Heidberg et al. (39) have deposited both platinum and iridium films from their respective acetylacetonates (acacs) by photoionizing the precursors with frequency-quadrupled radiation (266 nm) from an Nd:YAG (yttrium-aluminum-garnet) laser. The beam was passed 0.5 cm from the Si(111) substrate, resulting in the deposition of ~ 330-μm-wide metal stripes consisting of, for example, > 95 atom% Pt and the combined concentrations of C, O, and H: $< 5\%$. The metal acacs were vaporized into an Ar gas stream by heating the precursors with a pulsed CO_2 (10.6-μm) laser.

4.6. GROUP IB METALS (Cu, Au)

Despite the importance of the Column IB metals copper, silver, and gold to an array of fundamental microelectronics processes (doping, metallization, line and mask repair, etc.), the photodeposition of these metals from the vapor phase has been hindered by the scarcity of precursors. Simple alkyl molecules (i.e., of the form MX_n, where M = Cu, Ag or Au; X = CH_3, C_2H_5, C_4H_9, $\cdots$; and n is an integer) are not volatile, and the carbonyls $[\text{M(CO)}_n]$ are unavailable or do not exist. It was not until 1985 that Jones et al. (212) demonstrated the viability of volatile coordination complexes as precursors by depositing copper films upon photodissociating a fluorinated acetylacetonate complex, bis-(1,1,1,5,5,5-hexafluoropentanedionate)copper(II)–[Cu(Hfac)$_2$], with both pulsed and CW UV optical sources. A frequency-doubled Ar$^+$ ion laser (257 nm), ArF and KrF excimer lasers, and a low-pressure Hg lamp were all investigated.

Irradiation of Cu(Hfac)$_2$ with the Hg lamp yielded no deposition after 24 h, but the addition of ethanol to the vapor resulted in the production of a bright copper film on a quartz window in 3–5 h. No deposition was observed in experiments conducted at longer radiation wavelengths ($\lambda > 300$ nm), confirming that the process at 254 nm is primarily photochemical in origin. With the frequency-doubled Ar$^+$ ion laser, deposition was observed even in the absence of ethanol vapor with laser intensities of 0.2–2.0×10^4 W·cm^{-2} at a Si substrate. Films $\sim 1\,\mu$m in thickness were deposited at rates between 1 and 200 Å·min^{-1}. ArF or KrF laser irradiation of quartz substrates at peak intensities $\geqslant 1$ MW·cm^{-2} yielded thin Cu films (thicknesses $\leqslant 500$ Å), and the threshold intensity for film deposition was found to be ~ 1 MW·cm^{-2}. Regardless of the optical source, the resulting copper films also contained 10–90% carbon depending on whether ethanol was present or not. More detailed studies of the photodissociation of Cu(Hfac)$_2$ were subsequently carried out with a frequency-doubled Ar$^+$ (CW) laser by Houle, Wilson, and Baum (*213*), who showed that the degree of carbon incorporation into the copper films is determined by the intensity of the 257-nm radiation and the presence of ethanol vapor. The reduction in the carbon concentration that is observed upon introducing ethanol is shown in Table 12 (*213*). Small concentrations of oxygen and fluorine were also detected by AES. Note in Table 12 that the *relative* contributions of an element to the composition of both the precursors and the deposited films are indicated.

Analysis of the Cu films obtained by photolyzing Cu(Hfac)$_2$ at 257 nm showed (*213, 245*) that the relative concentrations of copper and carbon in the film were strongly dependent upon the laser intensity at the surface. Specifically, the surface morphology of the films was rippled and Auger microscopy revealed that the composition of the film varied dramatically across each of the ripples. As depicted in Figure 49, the copper composition peaks at the top of each ripple (where the UV intensity is the greatest) whereas the carbon signal is strongest in the valleys. These results suggest that the acetylacetonate groups on the Cu(Hfac)$_2$ precursor are photodissociated on the surface by the optical source and higher UV intensities are conducive to forming *volatile* products.

Similar comments could be made for the photodeposition of gold from a vapor phase precursor. Baum, Marinero, and Jones (*198*) demonstrated that gold can be patterned onto quartz substrates with a simple optical projection system to produce metal line widths as small as 2 μm. In their experiments, dimethylgold(III) acetylacetonate [Me$_2$Au(acac)] was photodissociated by

Table 12. Composition and Morphology of Films Photochemically Deposited from $Cu(Hfac)_2 \cdot (EtOH)_x$, $x = 0$–2

Light source	Precursor	Cu	C	O	F	Morphology
	$Cu(Hfac)_2$	1	10	4	12	—
Arc lamp[a]		—	—	—	—	No deposit
$Ar^+(SH)$[b]		1	4.3	—	—	Amorphous
Excimer[c]		1	68	16	15	Thin, discontinuous
	$Cu(Hfac)_2 \cdot EtOH$	1	12	5	12	—
Arc lamp		1	3.8	0.6	0.2	—
$Ar^+(SH)$		1	1.3	—	—	Amorphous
Excimer		1	27	1	4	Thin, discontinuous
	$Cu(Hfac)_2 \cdot (EtOH)_2$	1	14	6	12	—
Arc lamp		1	0.8	0.2	0.1	—
$Ar^+(SH)$		1	0.1	—	—	Amorphous
Excimer		1	4.5	3.6	—	Thin, discontinuous

Source: After Houle et al. (*213*).

[a] 254 nm.
[b] 257 nm. (The abbreviation *SH* denotes the second harmonic of the Ar ion laser at 257 nm.)
[c] 248 nm.

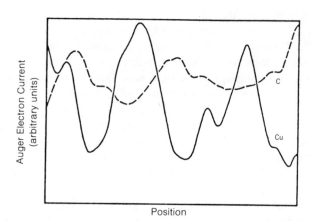

Figure 49. Relative copper and carbon compositions measured by an Auger scan across the surface of a film deposited by photolyzing $Cu(Hfac)_2$ at 257 nm (after ref. 245, by permission). The intermaxima spacing of the ripples is ~ 1800 Å, and the Cu concentration reaches its maximum value at the ripple peaks.

pulsed excimer radiation at 193 nm (ArF), 248 nm (KrF), or 308 nm (XeCl) or by an Hg resonance lamp. Films 2000–3000 Å thick were produced, and the highest gold content was achieved in films deposited with KrF laser radiation. Carbon content exceeded 50% in films deposited with 308- or 193-nm UV radiation. No deposition was observed if $\lambda = 351$ nm, which demonstrated that the film growth processes are photochemical in nature. Deposition was also observed with a Hg resonance lamp. Aylett and Haigh (199) also photo-deposited gold films by photodissociating trimethylgold(III)–trimethylphosphine $[(CH_3)_3Au–P(CH_3)_3]$ at 248 nm. Both broad-area and patterned deposition were demonstrated, the latter by projecting a mask onto the substrate.

Considerably more work has been reported in which copper or gold films were deposited from *liquids*. An example is the recent work of Baum (246) in which dimethyl (1,3-diphenyl-1,3-propanedionato)gold was photolyzed in solution by a 500-W Hg–Xe lamp. The resulting gold deposit acted as a catalyst for the spatially selective plating of Cu from an electroless bath. Details of similar processes and a discussion of the chemistry of photodeposition and etching in the liquid phase can be found in Haigh and Aylett (247) and Donohue (248).

4.7. SUMMARY

Of the metals that are of most interest to commercial processes (Au, W, etc.), many have been successfully photodeposited over broad areas (> 1 cm^2) and at competitive deposition rates by photo-CVD. However, carbon contamination is an ongoing problem for many photodeposited metals, since the precursors frequently involve CO or CH_3 groups. This is particularly true for the deposition of Cr, W, and Mo from their respective hexacarbonyls, in which dissociative chemisorption is a major factor. Two possible avenues to overcoming this difficulty are (1) to resort to other existing or new precursors, and (2) tailoring the gas background in which the precursor is entrained so as to promote the rapid removal of contaminant species from the surface. An example of the former is the movement away from TMA as a precursor for Al deposition and toward hydride- or amine-bearing polyatomics.

CHAPTER

5

SEMICONDUCTORS

Efforts to grow both elemental and compound semiconductor films by photo-chemical processes have accelerated dramatically over the last decade, and the growth of amorphous and epitaxial Si films have accounted for much of this activity. However, considerable success has also been realized in the deposition of II–VI and III–V compound films, and this chapter will summarize the results reported to date. Salient features of the deposition conditions and film properties can be found in Tables 13–16.

5.1. ELEMENTAL FILMS

5.1.1. Carbon

The deposition of amorphous carbon films has been accomplished by photo-dissociating halomethanes or vinylchloride (CCl_4, C_2H_3Cl) (*249*) or acetylene (*250*) with an ArF excimer laser. For C_2H_3Cl as a precursor, film deposition rates up to $150 \, \text{Å} \cdot \text{min}^{-1}$ were measured and the films contained up to ~ 1 atom% of chlorine (films deposited from CCl_4 consisted of up to 8 atom% Cl). Kitahama et al. (*250–253*) demonstrated that graphite films can also be deposited by the photodissociation of C_2H_2 in parallel geometry or when the substrate is irradiated at a glancing angle. Recently, Ohashi and co-workers (*62*) deposited amorphous hydrogenated carbon films by irradiating *n*-butane vapor with unfiltered radiation from a 750-MeV synchrotron (see Chapter 3 for details regarding the reactor design and experimental arrangement). Films having thicknesses $> 0.4 \, \mu\text{m}$ were deposited with *n*-butane pressures under 1 Torr, and the film deposition rate was found to increase linearly with negative bias on the substrate. Accordingly, the production of positive ions by the synchrotron radiation was identified as the dominant contributor to film deposition.

5.1.2. Silicon

All of the techniques that have been developed thus far for photodepositing Si may be classified into one of three categories: (1) Hg-photosensitized deposition; (2) direct decomposition of higher silanes (Si_2H_6, Si_3H_8); and (3) laser

97

Table 13. Summary of the Conditions under Which the Elemental (Column IV) Semiconductors Have Been Deposited

Element	Precursor	Optical Source	λ(nm)	Refs.	Experimental Conditions	Comments
C	$n\text{-}C_4H_{10}$	Synchrotron radiation (unfiltered)	Broadband (VUV and XUV)	62	750 MeV storage ring-integrated exposure of substrate ($\perp$ geometry): 500 mA·h; Si substrate, room temperature; substrate biased negatively with respect to Ni mesh anode; n-butane pressure < 1 Torr	Hydrogenated amorphous carbon films were deposited with thicknesses > 4000 Å; deposition rate (and film thickness) increased linearly with negative bias on the substrate. Dominant contribution to deposition due to positive molecular ions produced by photoionization of $n\text{-}C_4H_{10}$ by photons $\geqslant 30\,$eV in energy.
C	CCl_4, C_2H_3Cl	ArF laser	193	249	$\perp$ geometry; laser pulse energy: 60 mJ, 100 Hz repetition rate; 0.05–5.00 Torr gas pressure; mass flow rates: Ar, 100 sccm; precursor, 5–80 sccm; Si, glass and stainless steel substrates; $T \leqslant 400°C$	Deposition rates up to $\sim 150\,$Å·min^{-1} were observed for vinyl chloride at 1 Torr total pressure; films were amorphous carbon with 1–8 atom % chlorine.
C	C_2H_2	ArF	193	250	Laser energy: 50 mJ/pulse, 100 Hz; 1–30% acetylene in H_2; 8–75 Torr total pressure; Si substrate, 400–800°C	Films consisting of graphite layers were deposited in parallel geometry and in experiments in which the substrate was irradiated at an angle of 1°–30°.
Ge	GeH_4	ArF, KrF	193, 248	193	Room temperature quartz substrate; $\perp$ geometry; laser intensity: 1–11 MW·cm^{-2}, 20 Hz repetition frequency (< 2 W average power); GeH_4 partial pressure: 22.5 Torr (5% GeH_4 in He); exposure period: 420 s; mass flow rate ≈ 35 sccm	Polycrystalline films were deposited at rates up to 4–5 Å·s^{-1} for 193 nm excitation (2 MW·cm^{-2}), 5–6 Å·s^{-1} for 248 nm photons (8 MW·cm^{-2}); films consisted of randomly oriented, $\sim 0.5\,\mu$m grains; film thicknesses up to 1500 Å were obtained.
Ge	GeH_4	KrF	248	294	NaCl(100), quartz, and $1\bar{1}02$ sapphire substrates; temperature: 25–225°C (substrate heated with CW CO_2 laser); 10–40 mJ/pulse; 1.5–2.0 Torr GeH_4 in He	Epitaxial Ge grown on NaCl(100) at 120–150°C; on SiO_2, films are polycrystalline (0.3–0.7 μm grains); activation energy = 0.085 ± 0.020 eV.

98

Ge	GeH$_4$	193	ArF	‖ geometry: GaAs(100) substrate; $285 \leqslant T_s \leqslant 415^\circ$C; laser beam $\leqslant 2$ mm above substrate; 0.75 MW·cm^{-2} laser intensity, 40 Hz repetition frequency; 5% GeH$_4$ in He: total pressure = 5.6–30.4 Torr, 100 sccm mass flow rate; deposition period: 40–50 min; necessary for obtaining epitaxial films to etch substrate (*in situ*) with 5% HCl/He mixture (3.5 Torr, ∼ 50 sccm flow rate)	295, 296	Film growth rates of ∼ 6 to 50 Å·min^{-1} observed; switch to amorphous material occurs after film thickness reaches 400–700 Å; epitaxial films obtained at temperatures as low as 285°C.
Ge	GeH$_4$/NH$_3$	193	ArF	1 sccm NH$_3$ added to gas flow of refs. 295, 296 (above); laser pulse energy: 16.6 mJ·cm^{-2}, 40 Hz	140	Epitaxial film deposition rates up to ∼ 350 Å·min^{-1} were measured. Films were epitaxial for several hundred Å. Nitrogen in films was not detected by XPS or AES.
Ge, Si	GeH$_4$/Si$_2$H$_6$	193	ArF	‖ geometry (laser beam $\geqslant 1$ mm above substrate); Si(100) substrate, 250–350°C substrate temperature; Ge films: 140 sccm of 10% GeH$_4$/He at 50 Torr (laser: 60 Hz, 420 mW transmitted power); Si films: 40 sccm of 10% Si$_2$H$_6$ in He at 5 Torr (40 Hz, 200 mW transmitted power)	48	Amorphous Ge/Si structures were deposited with average Ge and Si layer thicknesses of 5.4 ± 0.2 and 10.7 ± 0.4 nm, respectively. Structures having up to nine layers were fabricated.
Ge/Si	GeH$_4$, Si$_2$H$_6$	193	ArF	Quartz and Si substrates, $T_S = 250$–500°C; laser pulse energy: 15 mJ·cm^{-2}, 40 Hz, beam $\geqslant 2$ mm above substrate; Si$_2$H$_6$/GeH$_4$ mixture (5% in He), 100 sccm, 10 Torr total pressure; Si$_2$H$_6$:GeH$_4$ partial pressure ratio variable	297	Alloy films were deposited for various compositions; deposition rates ranged from 150 to 300 Å·min^{-1}.

99

Table 13 (*Continued*)

Element	Precursor	Optical Source	λ (nm)	Experimental Conditions	Refs.	Comments
Ge	$C_2H_5GeH_3$, $(C_2H_5)_2GeH_2$	CO_2 laser	9.4, 10.6 μm	Laser intensity: 10–150 W·cm⁻² (CW); 80 Torr ethylgermane, 40 Torr diethylgermane; exposure period: 15–30 s	298	Polycrystalline films were deposited that could be doped by simultaneously photodissociating $Al_2(CH_3)_6$ or $Cd(CH_3)_6$; fragments resulting from photodissociation of the precursor were identified by their IR absorption spectra.
Si	SiH_4 (+ Hg vapor)	Hg lamp (low pressure)	185, 254	150–300°C substrate temperature; 1 Torr total pressure	255–258	Deposition rates of 35 Å·min⁻¹ were observed.
Si	SiH_4/B_2H_6 (+ Hg vapor)	Hg lamp	185, 254	Boron-doped amorphous hydrogenated films were deposited on glass (Corning 7059) substrates at temperatures ranging from 200 to 250°C; the ratio of the B_2H_6 to SiH_4 partial pressures was varied between $2.5 \cdot 10^{-5}$ and $1.0 \cdot 10^{-2}$	259	Film thicknesses were ~ 1 μm. It was concluded that the doping mechanism for photo-CVD films is the same as that for films prepared by glow discharge techniques.
Si	Si_2H_6	ArF, XeF	193, 351	Mass flow rates: Si_2H_6, 1 sccm; H_2, 9 sccm; total pressure: 0.15 Torr; substrate temperature: 550–750°C; Si(100) substrates; entrance window coated with FOMBLIN oil; ⊥ geometry; 100 Hz pulse repetition rate; 20 s exposure time; laser energy fluence: 40–60 mJ·cm⁻²	288	Epitaxial films were grown in 600–650°C range; above 700°C, deposition is by normal CVD. Below 700°C, activation energy is 1.7 eV for ArF irradiation and 1.6 eV for XeF. Without the UV, E_a was measured to be 1.95 eV. Deposition rate is tripled (over purely CVD rate) by laser. Deposition rates of 20 Å·min⁻¹ were recorded at 600°C with ArF *and* XeF irradiation of substrate.
Si	Si_2H_6	ArF	193	1% Si_2H_6 in He, total pressure = 10 Torr; 12 sccm mass flow rate; quartz substrate, temperature: 270°C; laser pulse energy: 22–70 mJ·cm⁻² per pulse, 10 Hz repetition frequency; metal mask with 21.5 or 130 μm stripe pattern	290	Patterned films were deposited on quartz; key process appears to be dissociative adsorption on substrate to form Si_2H_5 and H.

100

	Reactant	Source	Wavelength (nm)	Conditions	Ref.	Results
Si	Si_2H_6 (+ Hg vapor)	Hg lamp	185, 254	185 and 254 nm intensities available from lamp at 42 mm: 3.4 and 30 mW·cm⁻², respectively; substrate temperature: 100–300°C; inside of quartz optical window coated with vacuum oil to prevent deposition; Hg reservoir temperature: 0–40°C	264, 265	Amorphous Si film thicknesses up to 1 μm and film deposition rates up to 1300 Å·min⁻¹ were obtained for a substrate temperature of 250°C and an Hg reservoir temperature of 20°C. Disilane pressure: 1.5 Torr. Without Hg in the reactor, the deposition rate drops by 1–2 orders of magnitude. Deposition rate falls by ~40% in increasing the substrate temperature from 100 to 300°C (10 sccm Si_2H_6, 0.38 Torr partial pressure).
Si	SiH_4 (+ Hg vapor)	Hg lamp	185, 254	Hg lamp built into reactor to promote influence of 185 nm radiation on film growth	260	
Si	Si_2H_6/H_2, PH_3, B_2H_6 (+ Hg vapor)	Hg lamp	185, 254	Si_2H_6 flow rate: 1.5 sccm (n-type), 0.4 sccm (p-type); substrate temperature: 200°C; PH_3:Si_2H_6 ratio, 6000 ppm; B_2H_6:Si_2H_6 ratio, 10,000 ppm; lamp intensity: ~30 mW·cm⁻² at 3 cm; H_2:Si_2H_6 ratio, 40:350; total gas pressure: 2 Torr	266	n- and p-type a-Si:H and microcrystalline Si films were deposited at maximum rates of ~1 Å·s⁻¹ at 300°C substrate temperature. Dark conductivities of 20 S·cm⁻¹ for n-type and 1 S·cm⁻¹ for p-type films were measured. Optical band gaps ranged from 2.0 eV for n-type to 2.3 eV for p-type material.
Si	$Si_2H_6/SiH_2F_2/H_2$ (+ Hg vapor)	Hg lamp		Substrate temperature: 100–300°C; Si_2H_6, SiH_2F_2, and H_2 mass flow rates: 1, 20–30, and 150 sccm (respectively) at a total pressure of 2 Torr; lamp intensity: 30 mW·cm⁻²; Si(100) substrates	270	Epitaxial Si films were grown at temperatures down to 200°C at rates up to ~0.7 Å·s⁻¹. Film thicknesses of 1500–3000 Å were obtained, and the activation energy for the process was 0.18 eV. SiH_2F_2 flow was required to obtain epitaxial films.
Si	SiH_4/He (+ Hg vapor)	Hg lamp	185, 254	20% SiH_4 in He, 5 Torr total pressure; substrate temperature: 150–350°C; mass flow rates: 2–30 sccm; movable Teflon curtain installed to prevent deposition on window	269	Amorphous films were deposited at rates up to 1 Å·s⁻¹; optical band gap of films: 1.75 eV; pin solar cells fabricated from photo-CVD films yielded 8.5% efficiencies.

Table 13 (*Continued*)

Element	Precursor	Optical Source	λ(nm)	Experimental Conditions	Refs.	Comments
Si	Si_2H_6	Hg lamp	185, 254	1% Si_2H_6 in He and N_2 at atmospheric pressure; 400 sccm (Si_2H_6/He) and 500 sccm (N_2) mass flow rates; quartz and Si substrates, temperatures <400°C	273–275	~15 Å·min⁻¹ deposition rates were observed for ~80 mW·cm⁻² of incident lamp intensity. Growth rate roughly independent of temperature; amorphous films doped with B or P were also deposited; hydrogen content: 2–7 atom %.
Si	Si_2H_6	D_2 lamp	120–200	Si_2H_6 in N_2 carrier gas; lamp-heated Si(100) substrate, $550 \leq T_s \leq 650$°C; Si_2H_6 mass flow rate: 1.5–15 sccm; total gas pressure: 0.17 Torr	279	Epitaxial films were deposited at 650°C at a rate of ~900 Å·min⁻¹; activation energy over 550–650°C range: 1 eV; VUV cleaning of substrate was also demonstrated.
Si	Si_2H_6/XeF_2	Hg lamp	185, 254	Lamp intensity at 3 cm: 7 mW·cm⁻² at 185 nm; Si and glass substrates; 245–350°C substrate temperature; XeF_2 mass flow rate: 0.03–0.95 sccm (1.5–45 mTorr pressure); Si_2H_6 and He partial pressures: 0.3–2.3 and 0.7–2.7 Torr, respectively	180	Amorphous hydrogenated Si films were deposited using XeF_2 as an etchant to keep the reactor entrance window free of deposition. Deposition rates up to 36 Å·min⁻¹ were obtained for a XeF_2 mass flow rate of ~1 sccm and Si_2H_6 pressure of 2.3 Torr.
Si	Si_2H_6	D_2, Xe lamps	120–200, 147	150 W D_2 lamp, microwave-excited Xe lamp; Si_2H_6 only (no carrier), 2.5 sccm mass flow rate, 2.5 Torr total pressure; Si and glass substrates at temperatures between 200 and 300°C	277	Deposition rates of ~15 Å·min⁻¹ for the D_2 lamp and up to 75 Å·min⁻¹ for the Xe lamp (at ~120 W input power to lamp) were measured; no dependence of deposition rate on temperature for D_2 lamp in 200–300°C range.

			Wavelength (nm)	Conditions	Ref.	Comments
Si	Si_2H_6	Xe lamp (pulsed)		Lamp parameters: $\sim 110\,mJ$ input energy per pulse, several μs pulse width, 50 Hz repetition frequency; 10% Si_2H_6 in He, 50 sccm mass flow rate; substrate temperature: $< 300°C$	278	Early stages of nucleation were explored; only $\lambda < 200\,nm$ photons are effective in dissociating Si_2H_6 precursor; between 150 and 250°C, process activation energy $= 0.15\,eV$.
Si	Si_2H_6	Windowless He lamp	121.5	20 sccm of Si_2H_6 and 200 sccm of He; total pressure: 1–1.5 Torr; glass (Corning 7059), quartz and Si(100) substrates; $50 < T_s < 400°C$	175	a-Si:H films were deposited at rates $> 200\,\text{Å} \cdot min^{-1}$; films had an optical band gap of 1.76 eV; hydrogen concentrations of 4.3% and photoconductivity (AM1) of $4 \cdot 10^{-4}\,(\Omega \cdot cm)^{-1}$.
Si	Si_3H_8	D_2 lamp		$2\%\ Si_3H_8$ in H_2 (6.7–47 sccm Si_3H_8, 330–4100 sccm H_2); substrate temperature: 500–650°C; quartz substrate; deposition time: 2.5–20 min	281	Polycrystalline films were deposited with thicknesses between 2.5 and 10 μm; strong $\langle 110 \rangle$ or $\langle 100 \rangle$ preferred orientation.
Si	SiH_4/N_2	ArF, KrF	193, 248	Room temperature quartz substrate, $\perp$ geometry; laser intensity: 1–20 MW·cm^{-2}, 20 Hz repetition frequency	193	Maximum Si growth rate of $9.8\,\text{Å} \cdot s^{-1}$ observed at 248 nm and 20 MW·cm^{-2} peak laser intensity; polycrystalline films up to 4000 Å in thickness were deposited.
Si	Si_2H_6, Si_3H_8	ArF	193	Glass, quartz, and Si(100) substrates; temperatures: 35–350°C; $\perp$ geometry; 10 Torr Si_2H_6	284–286	Amorphous hydrogenated Si films were deposited at rates ranging from 100–300 $\text{Å} \cdot min^{-1}$ (see Table 14).

Table 14. Properties of Undoped a-Si:H Films Photodeposited with Different Optical Sources

σ_{ph} (AM1)$(\Omega\cdot cm)^{-1}$	1.2×10^{-5}	2×10^{-4}	2×10^{-6}	10^{-5}	6×10^{-4}	4×10^{-4}	1×10^{-3}	$\sim 10^{-4}$
σ_D	2.5×10^{-11}	2×10^{-9}	1×10^{-10}	10^{-10}	4×10^{-9}	3.5×10^{-9}	3×10^{-9}	$\sim 10^{-11}$
Donor gas	Si_2H_6	Si_2H_6	Si_2H_6	Si_2H_6, Si_3H_8	Si_2H_6	Si_2H_6	Si_2H_6	Si_2H_6
Energy source	193 nm excimer laser	193 nm excimer laser	193 nm excimer laser	Hg lamp: 184.9 nm	Hg lamp: 254 nm + weak 185 nm (no Hg in reactor)	He$^+$ lamp: 121.5 nm He II line & atomic helium metastables	RF Plasma	H$_2$ lamp (internal)
Deposition rate	200–300 Å·min^{-1}	120 Å·min^{-1}	100–200 Å·min^{-1}	6 Å·min^{-1}	15 Å·min^{-1}	200 Å·min^{-1}	—	240 Å·min^{-1}
Reference	223	243	218	216	199	143	176	272

Source: Adapted from Zarnani et al. (175).

104

Table 15. Summary of II–VI Compound Films Deposited by Photochemical Processes

Element	Precursor	Optical Source	λ(nm)	Experimental Conditions	Refs.	Comments
CdTe	$Cd(CH_3)_2$, $Te(C_2H_5)_2$	Hg lamp	185, 254	GaAs(100), CdTe or quartz substrates; substrate temperature: 250–420°C; partial pressures of organometallic precursors: 0.08 Torr; total pressure: 1 atm; total mass flow rate: 1000 sccm	313, 316, 317	Epitaxial films up to 3 μm thick were obtained for temperatures as low as 250°C; obtaining crystalline films requires the Cd:Te precursor concentration ratio to be 1.5:1.0; film growth rates up to 0.9 μm·h^{-1} at 250°C and 6 μm·h^{-1} at 350°C were measured.
CdTe	$Cd(CH_3)_2$, $Te(C_2H_5)_2$, or $Te(C_4H_9)_2$	Hg/Xe arc lamp (1 kW)		200–250 nm radiation from lamp is directed onto substrate by a dichroic mirror; CdTe, ZnCdTe, and Si substrates; substrate temperature: 182 and 240°C	173, 314	CdTe films were deposited to fabricate HgTe–CdTe superlattices.
CdTe	$Cd(CH_3)_2$, $Te(C_2H_5)_2$	ArF, KrF lasers	193, 248	Undoped GaAs(100) substrates; laser fluence: 10–120 mJ·cm^{-2}, 10–80 Hz repetition frequency; mass flow rates and partial pressures: 10^{-3}–10^{-2} Torr and 36 sccm [$Cd(CH_3)_2$]; 10^{-3}–10^{-1} Torr, 178 sccm [$Te(C_2H_5)_2$]; total pressure: 10–80 Torr; substrate temperature: 165°C; $\perp$ or $\parallel$ geometry	37, 179, 324	Growth rates up to 2 μm·h^{-1} were realized for epitaxial CdTe(111) on GaAs at 165°C; high-quality films are deposited in parallel geometry.
CdTe	$Cd(CH_3)_2$, $Te(C_2H_5)_2$	ArF, KrF	193, 248	$\parallel$ geometry; 100–150°C growth temperatures; CdTe substrates; 7.5–15 mJ/pulse, 200–500 Hz repetition frequency; total pressure: 10–100 Torr	38	Deposition rates up to ~1 μm·h^{-1} were obtained and surfaces were specular.

Table 15 (*Continued*)

Element	Precursor	Optical Source	λ (nm)	Experimental Conditions	Refs.	Comments
CdTe	Cd(CH$_3$)$_2$, Te(C$_2$H$_5$)$_2$	Hg lamp (2 kW)		Lamp was microwave excited, yielding an average intensity in the 200–250 nm range of 100 mW·cm^{-2}; GaAs(100) and (111) substrates; $\perp$ geometry; 250°C substrate temperature; dimethylcadmium and diethyltellurium partial pressures: 5.4–7.4·10^{-4} and 1.6·10^{-3} atm, respectively (0.41–0.56 Torr and 1.2 Torr, respectively); overall pressure—atmospheric; growth times: 10 min–1.5 h	318,319	Maximum growth rates were measured to be 13 and 9 μm·h^{-1} for CdTe and GaAs(100) substrates, respectively. The deposition rate was found to be linear in optical source (lamp) intensity.
Cd$_x$Hg$_{1-x}$Te	Cd(CH$_3$)$_2$, Te(C$_2$H$_5$)$_2$, Hg vapor	Hg lamp (3 kW)		CdTe(100), (110), or InSb substrates; substrate temperature: 250°C; 1 atm total pressure (He carrier gas); Hg and Cd(CH$_3$)$_2$ partial pressures: ~ 23 and 4.5 Torr, respectively	313,321,322	Obtaining epitaxial films requires the use of He rather than H$_2$ as the carrier gas to suppress nucleation in the vapor. Epitaxial films 1.3 μm thick were grown at 250°C; Cd$_x$Hg$_{1-x}$Te/CdTe interfaces < 400 Å in width were obtained at 250°C.
Cd$_x$Hg$_{1-x}$Te	Cd(CH$_3$)$_2$, Te(CH$_3$)$_2$, Hg(CH$_3$)$_2$	ArF laser	193	Cd(CH$_3$)$_2$, Hg(CH$_3$)$_2$, and Te(CH$_3$)$_2$ mass flow rates of 1, 10, and 5 sccm, respectively; He mass flow rate: 64 sccm; total pressure: 5 Torr; substrate: CdTe at 150°C; exposure time: 30 min; parallel geometry; ~ 380 mJ/pulse laser energy	323	Growth rates of 4 μm·h^{-1} (2.25 μm thick film grown in 30 min) were obtained. Film stoichiometry: Cd$_{0.8}$Hg$_{0.2}$Te.
Cd$_x$Hg$_{1-x}$Te	Cd(CH$_3$)$_2$, Te(C$_2$H$_5$)$_2$ or methylallyltelluride, Hg vapor	Hg lamp (2 kW)		Lamp was microwave excited with an average intensity of 100 mW·cm^{-2} in the 200–250 nm region; substrate temperature: 250°C; Cd(CH$_3$)$_2$, Te(C$_2$H$_5$)$_2$, and Hg	319	4 μm·h^{-1} photodeposition rates were found for growth of Cd$_{0.2}$Hg$_{0.8}$Te at 250°C when methylallyltelluride was the Te precursor; variation in x (Cd$_x$Hg$_{1-x}$Te)

Material	Precursor	Light source	Ref.	Conditions	Ref.	Results
				vapor partial pressures: $5.8 \cdot 10^{-4}$, $1.9 \cdot 10^{-3}$, and 0.03 atm, respectively (~ 0.4, 1.4, and 22.8 Torr, respectively); $Cd_xHg_{1-x}Te$ films were grown on photo-CVD CdTe layers grown on GaAs/Si substrates		was 4% over 1 cm² of area.
HgTe	$Te(C_2H_5)_2$, Hg vapor	Hg arc lamp	185, 254	Films deposited by Hg photosensitization; substrate temperature: 180–310°C; InSb and CdTe substrates; $Te(C_2H_5)_2$ partial pressure: $\sim 5 \cdot 10^{-3}$ atm (~ 4 Torr); $4 \cdot 10^{-2}$ atm (~ 30 Torr) (250°C); 1 atm total pressure	310, 313	Epitaxial film growth rates of 1–2 $\mu m \cdot h^{-1}$ were obtained in the 240–310°C region. At 200°C, deposition rate is ~ 0.4–$0.6\ \mu m \cdot h^{-1}$; at 180°C: $0.1\ \mu m \cdot h^{-1}$; films were $\sim 1\ \mu m$ thick.
HgTe	$Te(C_2H_5)_2$ or $Te(C_4H_9)_2$, Hg vapor	Hg/Xe arc lamp (1 kW)	193, 248	200–250 nm radiation from lamp is directed onto substrate by a dichroic mirror; CdTe, ZnCdTe, and Si substrates; substrate temperature: 182 and 240°C	173, 314	HgTe–CdTe superlattices were deposited on CdTe with growth rates of $< 0.1\ \mu m \cdot h^{-1}$ at 182°C for the diethyl precursor and $< 0.25\ \mu m \cdot h^{-1}$ for the diisopropyl precursor. At 240°C, the deposition rate is $\sim 1\ \mu m \cdot h^{-1}$ for $Te(C_4H_9)_2$.
HgTe	$Te(C_2H_5)_2$, Hg vapor	ArF, KrF		$\parallel$ geometry: 100–150°C growth temperature; CdTe substrate; 7.5–15 mJ/pulse, 200–500 Hz repetition frequency; total pressure: 10–100 Torr	38	Deposition rates up to $\sim 1\ \mu m \cdot h^{-1}$ were obtained; epitaxial films were grown on CdTe(111), and surfaces were specular.
HgTe	$Te(C_2H_5)_2$, $Hg(CH_3)_2$, or $Hg(C_2H_3)_2$	KrF laser	248	GaAs(100) substrate: 165°C; 10–120 mJ·cm⁻² laser energy fluence; partial pressures: 10^{-3}–10^{-1} Torr dimethyltellurium; 10^{-2}–1 Torr dimethylmercury, 10^{-2}–10^{-1} Torr divinylmercury; 10–80 Torr total pressure (He carrier)	324	Epitaxial films were grown at 165°C at a rate of $0.8\ \mu m \cdot h^{-1}$ (15 min exposure period); the laser fluence was 60 mJ·cm⁻², and better film stoichiometry was obtained with the divinyl precursor.

Table 15 (*Continued*)

Element	Precursor	Optical Source	λ (nm)	Experimental Conditions	Refs.	Comments
ZnO	$Zn(C_2H_5)_2$, N_2O (or NO_2)	ArF, KrF lasers	193, 248	$\parallel$ and $\perp$ irradiation geometries; Si substrates; substrate temperature: room temperature to 220°C; best results with NO_2 mass flow rate of 34 sccm and $Zn(C_2H_5)_2$ partial pressure of 30 mTorr; total pressure of 2 Torr	178	Deposition rates up to 3000 Å·min⁻¹ were measured; $\perp$ geometry yielded highest optical quality and lowest resistivity films. Thickness uniformity $\pm 5\%$ over 2 cm × 5 cm area; ESCA measurements show film composition to be 49% Zn, 51% oxygen.
ZnS	$Zn(CH_3)_2$, $S(C_2H_5)_2$	ArF, KrF	193, 248	$\parallel$ geometry, 100–150°C growth temperature, GaAs(100) substrate; 7.5–15 mJ/pulse, 200–500 Hz repetition frequency; total pressure: 10–100 Torr; H_2 carrier gas mass flow rate: 500 sccm	38	Deposition rates up to $\sim 1\ \mu m \cdot h^{-1}$ were measured; surfaces were specular in growth temperature range investigated.
ZnS	$Zn(CH_3)_2$, $S(C_2H_5)_2$, or CH_3SH	Xe arc lamp (500 W)		Lamp intensity: 42 mW·cm⁻² at substrate [GaAs(100)]; 275–600°C temperature; (0.5–4.0)·10⁻⁵ mol·min⁻¹ dimethylzinc; 8·10⁻⁵ mol·min⁻¹ diethylsulfide, 4000 sccm H_2; 760 Torr total pressure	303	Deposition rates > 3 $\mu m \cdot h^{-1}$ observed at 400°C with irradiation; growth of epitaxial ZnS and ZnSe films at low temperature was shown to be associated with photogenerated carriers.

108

Material	Precursors	Light source	Wavelength	Ref.	Conditions	Comments
ZnSe	$Zn(CH_3)_2$, $Se(CH_3)_2$	Hg/Xe arc lamp (1 kW)		225	Quartz substrates; lamp power for $\lambda < 240$ nm: ~1 W; 1–2 h deposition times; 1–10 Torr $Zn(CH_3)_2$ and $Se(CH_3)_2$, 0–620 Torr Ar; deposition at room temperature	Film thicknesses up to ~0.6 μm were obtained over >3 cm² of substrate; rough film morphology.
ZnSe	$Zn(CH_3)_2$, $Se(CH_3)_2$	Xe arc lamp (1 kW)		301–303	Maximum lamp intensity at substrate: $100\,mW\cdot cm^{-2}$; both ∥ & ⊥ configurations; GaAs(100) substrates; temperature: 300–500°C; flow rates: dimethylzinc, $10\,\mu mol\cdot min^{-1}$; dimethylselenide, $40\,\mu mol\cdot min^{-1}$; H_2; 1000 sccm; total pressure: 200 Torr	At 350°C, the deposition rate with substrate irradiation ($\perp$, $47\,mW\cdot cm^{-2}$) is $1.5\,\mu m\cdot h^{-1}$; without irradiation: $<0.04\,\mu m\cdot h^{-1}$. At 400°C, growth rate increases linearly with lamp intensity up to $\sim 40\,mW\cdot cm^{-2}$. Epitaxial layers were obtained.
ZnSe	$Zn(CH_3)_2$, $Se(C_2H_5)_2$	ArF	193	309	∥ geometry: GaAs(100) substrate; 100–575°C substrate temperature; 2.5 sccm dimethylzinc, 25 sccm diethylselenide, 2000 sccm H_2; laser pulse energy: 50 mJ, 20 Hz repetition frequency	Epitaxial ZnSe films were grown at rates of $\sim 1\,\mu m\cdot h^{-1}$ at temperatures as low as 200°C.
ZnSe	$Zn(C_2H_5)_2$, $Se(CH_3)_2$	Hg lamp	185, 254	234	Lamp intensity: $40\,mW\cdot cm^{-2}$; total pressure in reactor: ~1 Torr	Epitaxial ZnSe films were deposited at temperatures down to 200°C.

Table 16. Summary of Conditions under Which III–V Semiconductor Compound Films Have been Grown by Photo-CVD

Compound	Precursors	Optical Source	λ (nm)	Experimental Conditions	Refs.	Comments
GaAs	Ga(CH₃)₃, AsH₃	Hg lamp, ArF, XeF lasers	185, 254, 193, 351	GaAs(100) substrates heated to 510–640°C by a quartz halogen lamp; H₂ carrier gas; lamp intensity at substrate: 0.6 mW·cm⁻², excimer laser intensity: 2.8 MW·cm⁻², 10 Hz (193 nm), 2.4 MW·cm⁻², 10 Hz (351 nm)	325, 326	Maximum film deposition rate increased from 5 to 8 μm·h⁻¹ with lamp irradiation of substrate at 510°C. Smaller improvements were measured at higher substrate temperatures (600, 650°C); Mirror-like surfaces were obtained at 550°C with UV lamp radiation; with ArF radiation, the deposition rate doubled but no effect was observed at 351 nm.
GaAs	GaCl₃, AsCl₃	ArF, KrCl, KrF, XeCl, XeF lasers	193, 222, 248, 308, 351	GaAs(100) substrates—offcut 4° toward (110). Substrate temperature: 350–700°C; ⊥ geometry; laser repetition frequency: 5–70 Hz	327–329	Film growth rate was enhanced significantly over 480–700°C range with KrF laser irradiation; epitaxial films were obtained at temperatures as low as 350°C; For a substrate temperature of 600°C, GaAs growth rates up to 4.5 μm·h⁻¹ were observed for 2.7 W of average 248 nm laser power—this deposition rate is a factor of ~4 improvement over that observed with no UV irradiation of the substrate. Laser radiation at 248 or 222 nm promotes the reduction of AsCl₃ by H₂.
GaAs	Ga(CH₃)₃ or Ga(C₂H₅)₃, AsH₃	ArF, KrCl, KrF, XeCl, XeF lasers	193, 222, 248, 308 351	Semi-insulating or Si-doped GaAs substrates; ⊥ geometry; 60 Hz laser repetition frequency	330	UV laser enhanced surface migration of adsorbed species, permitting crystalline films to be grown at temperatures as low as 350°C for Ga(CH₃)₃ and 300°C for Ga(C₂H₅)₃; Hall mobility also increased with UV photons at the surface.

Material	Precursors	Light source	Ref./wavelength	Conditions	Results	Ref.
GaAs	$Ga(CH_3)_3$, AsH_3 (H_2 carrier)	ArF laser	193	n-Type GaAs(100) substrates; substrate temperature: 425–500°C; AsH_3:$Ga(CH_3)_3$ molar ratio: 4–30; total reactor pressure: 10–30 Torr; laser pulse energy: 50–100 mJ ($\sim$19–38 mJ·cm^{-2}; 40–80 Hz repetition frequency; $\perp$ geometry	Epitaxial films were grown at temperatures as low as 425°C with deposition rates ranging from 0.02 μm·min^{-1} at 425°C to 0.1 μm·min^{-1} at 500°C (laser: 70 Hz, 36 mJ/cm^2). Film properties are similar to those of layers grown by conventional MOCVD (hole concentrations: $5\cdot10^{16}$–$3\cdot10^{17}$ cm^{-3}; hole mobilities: 150–200 cm^2/V·s); carbon concentrations: $4\cdot10^{17}$–$2\cdot10^{18}$ cm^{-3} (500°C growth temperature).	332
GaAs	$Ga(CH_3)_3$, AsH_3 (H_2 carrier)	ArF, XeF	193, 351	Si-doped GaAs(100) substrates; substrate temperature: 450–500°C; $\perp$ geometry; mass flow rates: 30–50 sccm (AsH_3) and 3 sccm ($Ga(CH_3)_3$); laser energy fluence: $\leqslant$ 13 mJ·cm^{-2} at 30 Hz	Film deposition rates at 450°C were enhanced by 5–15% with ArF irradiation which was attributed to photolysis of adsorbed trimethylgallium. No effect was observed for XeF photons.	333
GaAs	$Ga(CH_3)_3$, AsH_3 (H_2 carrier)	Hg lamp (high pressure), CO_2 laser	$\geqslant$ 300; 9.26–10.53 μm (CO_2)	GaAs[(100), 2° →(110)] substrates, heated by halogen lamp; substrate temperature: 400–700°C; total mass flow rate: 203 sccm; CO_2 laser tuned to 10.531 μm absorption line of AsH_3; CO_2 intensity (CW): 11 W·cm^{-2}	CO_2 laser enhanced growth rate from $\sim$ 1.5 to 4 μm·h^{-1} at 500°C, and Hg lamp improved film quality; effect of CO_2 laser attributed to AsH_3 vibrational excitation at surface. No CO_2 laser effect was observed at 9.26 μm, where AsH_3 does not absorb.	337
GaAs	$Ga(C_2H_5)_3$, AsH_3 (N_2, H_2 carriers)	Hg–Xe lamp (1000 W), Hg grid lamp (500 W)		Quartz substrates, $T = 240$°C; mass flow rates: 5 sccm [$Ga(C_2H_5)_3$], 62 sccm (10% AsH_3 in H_2); total pressure: 2 Torr; 60 min growth period	1.6 μm thick polycrystalline films were deposited; no deposition occurs in the absence of lamp radiation; deposition by Hg photosensitization was also demonstrated with the grid lamp.	338, 339

111

Table 16 (*Continued*)

Compound	Precursors	Optical Source	λ (nm)	Experimental Conditions	Refs.	Comments
GaAs, AlGaAs	Ga(CH$_3$)$_3$, Al(CH$_3$)$_3$, AsH$_3$ (H$_2$ carrier)	ArF laser	193	Cr-doped GaAs(100) substrates; substrate temperatures: 550–750°C (GaAs), 570–800°C (AlGaAs); mass flow rates: 0.02–0.14 sccm Al(CH$_3$)$_3$; 0.4–1.0 sccm Ga(CH$_3$)$_3$ [for GaAs], 0.12–0.36 sccm [for AlGaAs]; 2–200 sccm AsH$_3$; total pressure: 50–200 Torr; ArF laser energy fluence: 30 mJ·cm^{-2}; 50 Hz repetition frequency (1.5 W·cm^{-2})	331	Carrier concentrations increase in n-type regions and decrease in p-type regions with irradiation (at 100 Torr reactor pressure); aluminum content in AlGaAs rises with UV illumination, but growth rates are not affected by 193 nm photons for $T > 600$°C.
GaAlAs, AlAs	Al(C$_4$H$_9$)$_3$, As vapor, Ga(C$_2$H$_5$)$_3$	ArF laser	193	GaAs substrate; substrate temperature: 350°C; precursor partial pressures: $1 \cdot 10^{-6}$ [Ga(C$_2$H$_5$)$_3$], $1 \cdot 10^{-5}$ [Al(C$_4$H$_9$)$_3$] and $5 \cdot 10^{-5}$ Torr (As$_4$ from elemental arsenic; for AlAs growth, TIBA pressure was reduced to $5 \cdot 10^{-6}$ Torr; laser energy: 60 mJ/pulse, 50 Hz repetition frequency	80	Epitaxial AlAs films up to 1.25 μm in thickness were grown at 350°C (1 h growth time); without ArF irradiation, films were ~ 500 Å in thickness; poly-crystalline GaAlAs films 0.2 μm in thickness were grown at 350°C; without laser, films were 200–300 Å thick; Al content was higher in irradiated areas.
GaAs, AlAs, [Al, Ga]-As	(CH$_3$)$_3$Ga: As(CH$_3$)$_3$, and (CH$_3$)$_3$Al:As (CH$_3$)$_3$ adducts	ArF, KrF lasers	193, 248	Ge(100) substrates; 20–700°C substrate temperature; adduct partial pressures: 10^{-3}–0.4 Torr; 30 min growth period; excimer laser fluence: 20–300 mJ·cm^{-2}, 10 Hz repetition frequency; $\perp$ and $\parallel$ geometries	354	GaAs film quality was highest when 193 nm laser beam is parallel to substrate; for 248 nm fluences above 140 mJ·cm^{-2}, As/Ga ratio increases ($\perp$ geometry); depositions at 193 nm were carried out at 300°C; highest growth rates measured were 0.03 Å/pulse.

112

Material	Precursors	Laser	λ (nm)	Conditions	Ref.	Description
GaN	Ga(CH$_3$)$_3$, NH$_3$	ArF laser	193	GaAs(100) and sapphire (0001) [basal plane] substrates; substrate temperature: 600–900°C; ∥ geometry	358	Polycrystalline films preferentially oriented (0001) were grown on basal plane sapphire at 625°C at rates exceeding 2 μm·h^{-1}.
GaP	Ga(CH$_3$)$_3$, tert-butylphos-phine (TBP)	ArF laser	193	Sulfur-doped, GaP(100) substrates; substrate temperature: 500°C; mass flow rates: 3 (TMG) and 24 sccm (TBP); total gas pressure: 30 Torr; laser fluence: 60–120 mJ·cm^{-2}, 50 Hz repetition rate	355–357	Epitaxial GaP films were grown at 500°C; at lower temperatures and in the absence of UV laser radiation, films are polycrystalline. Obtaining epitaxial films requires ~100 mJ·cm^{-2} of fluence. Beyond 120 mJ·cm^{-2}, sputtering dominates. Epitaxial films could not be grown at $\lambda = 248$ nm.
InP	(CH$_3$)$_3$InP(CH$_3$)$_3$, P(CH$_3$)$_3$	ArF, KrF, XeF lasers	193, 248, 351	Quartz, GaAs, and InP substrates; temperatures up to 400°C; mass flow rates: (CH$_3$)$_3$InP(CH$_3$)$_3$: ≤ 33 sccm, P(CH$_3$)$_3$: 36 sccm, He: 230 sccm; total pressure: 2.1 Torr; ~10 mJ/pulse laser energy, 125 mJ·cm^{-2} fluence, 8 MW·cm^{-2} peak intensity, 5 Hz repetition frequency; ⊥ geometry; exposure: 4200 pulses	104, 359, 360	Amorphous, polycrystalline, and epitaxial InP films were deposited. Epitaxial films were grown at ~320°C and laser fluences (⊥ geometry) of 100 mJ·cm^{-2}; deposition rates of ~0.2 Å/pulse were obtained.
InSb	In(CH$_3$)$_3$, Sb(CH$_3$)$_3$	ArF, KrF	193, 248	Quartz, GaAs(100), and InSb(100) substrates at room temperature; precursors diluted in H$_2$; total reactor pressure: 620 Torr; laser fluence: 50 mJ·cm^{-2}, 4 Hz repetition frequency; exposure: 5000 pulses	35	Polycrystalline InSb films were deposited on GaAs at room temperature. The highest quality films were deposited with KrF radiation (248 nm) and a Sb(CH$_3$)$_3$: In(CH$_3$)$_3$ ratio of 17:1.

photodeposition where the first two approaches generally rely on lamps as the optical source.

5.1.2.1. Mercury-Photosensitized Deposition

Since the pioneering work of Emeléus and Stewart (2), Niki and Mains (3), and Rousseau and Mains (254) exploring the mercury-photosensitized decomposition of monosilane (SiH_4) and germane (GeH_4), amorphous hydrogenated Si (a-Si:H) films, microcrystalline and epitaxial Si films have been deposited by a number of groups. Saitoh et al. (255–257) and Tarui and co-workers (258) deposited thin films of amorphous or polycrystalline Si by the Hg-photosensitized decomposition of monosilane. These studies showed photo-CVD deposited a-Si films to be of comparable quality to plasma-CVD-deposited films with the added benefit that ion induced damage is not present in photochemical processing. Recently, Dutta et al. (259) compared the electrical and structural properties of microcrystalline hydrogenated Si (μc-Si:H) films grown by Hg-sensitized photo-CVD and glow discharge. All of the films were grown in commercial reactors and, based on dark conductivities, grain size, and preferred microcrystallite orientation, the photodeposited films were judged to be superior to their discharge-deposited counterparts for photovoltaic device applications.

Early work in this area reported film deposition rates of 20–40 Å·min^{-1}, which were subsequently increased to 600 Å·min^{-1} by improvements in reactor and Hg lamp design. In the latter area, Aota et al. (260) built an Hg lamp directly into the reactor to promote the participation of 185-nm resonance radiation in the film deposition process.

Mizukawa and co-workers (261) deposited p-type amorphous Si films by adding small concentrations of diborane (B_2H_6) to the silane vapor in the reactor (200–250°C substrate temperatures). Similar studies conducted by Suzuki et al. (262) involved doping microcrystalline silicon by adding either PH_3 or B_2H_6 to the $SiH_4/H_2/Hg$ feedstock gas stream. The conductivity of the μc-Si:H films was found to be more than 2 orders of magnitude greater than that for a-Si:H films. For a PH_3/SiH_4 or B_2H_6/SiH_4 ratio of $\sim 2 \times 10^{-4}$, the film deposition rate was 150 Å·min^{-1} and films were grown at substrate temperatures as low as 150°C.

Kinetic studies of the Hg photosensitization kinetics chain (263) demonstrated that the rate of production of Si_2H_6 (disilane) and Si_3H_8 (trisilane), which are a direct result of the collisional dissociation of SiH_4, scale linearly with the source intensity. In an effort to bypass the unnecessary conversion of SiH_4 to Si_2H_6, most recent Si photodeposition by Hg sensitization has utilized Si_2H_6 directly. Inoue, Konagai, and Takahashi (264), and Tarui et al. (265) reported the deposition of amorphous Si films having thicknesses up to ~ 1 μm by irradiating an Si_2H_6/Hg mixture for 10 min. Without the presence of Hg in the reactor, the film deposition rate was observed to fall by 1–2 orders of magnitude. Both n- and p-type amorphous hydrogenated and microcrystalline Si films have also been deposited (266) by adding phosphine (PH_3) or B_2H_6, respectively, to the reactor. The ratio of PH_3 or B_2H_6 to Si_2H_6 was 6000 or 10,000 ppm, respectively, and the total gas pressure in the reactor was 2 Torr. Solar cells (SiC/Si $p-i-n$ structures) have also been fabricated (267–269) from photodeposited amorphous films and have yielded efficiencies at AM1.5 (air mass 1.5*) of up to 8.5% (270).

A major development in the application of Hg photosensitization to film deposition was the report of Nishida et al. (270) in 1986 that epitaxial Si films could be grown at substrate temperatures as low as 200°C by this process. The addition of SiH_2F_2 to the gas stream was found to be critical, not only in the early stages of film growth, but throughout the deposition process. It was concluded that halogen-bearing radicals or free F atoms may be responsible for removing the native oxide. Mass flow rates for SiH_2F_2 exceeding ~ 20 sccm (standard cc per minute) at a total reactor pressure of 2 Torr were necessary to grow epitaxial films and deposition rates up to $0.7 \text{Å} \cdot \text{s}^{-1}$ were observed. The activation energy[†] for this low-temperature process was measured to be 0.18 eV, which is similar to the value of 0.09 eV reported in 1985 by Osmundsen et al. (101) for the KrF laser-initiated growth of Ge films. If D_2 is substituted for H_2 as the carrier gas, Jia et al. (271) found that the maximum substrate temperature at which crystalline Si films could be grown by the photosensitization of $SiH_4/SiH_2F_2/H_2$ (or D_2) mixtures was raised to 475°C. Crystallinity and surface morphology of the films improved significantly, and Figure 50 shows the variation of the electron mobility and resistivity with carrier concentration for Si films grown at 300°C and doped with phosphorus (from PH_3). The bulk values for each parameter are represented in the figure by the dashed curves.

5.1.2.2. Direct Photodecomposition of SiH_4, Si_2H_6, or Si_3H_8

The underlying motivation behind Hg-photosensitized deposition of Si from monosilane is the weak absorption of the precursor in the UV and at the 185 and 254 nm lines of the Hg resonance lamp, in particular. Because of the increasing absorption at longer wavelengths that is associated with resorting to progressively higher silanes (i.e., $SiH_4 \rightarrow Si_2H_6 \rightarrow Si_3H_8$), however, one can avoid the presence of Hg in the reactor entirely by photodissociating disilane (or trisilane) directly.

---

*AM1 (air mass 1) is the equivalent of the sun's intensity at the earth's surface.

[†]The dependence of film growth rate on temperature can be expressed as $R \propto R_0 \exp(-E_a/kT)$, where E_a is known as the activation energy. An earmark of photo-CVD is the weakened coupling between growth rate and temperature. This is manifested by E_a's that, for a given material, are generally smaller for photochemical growth than for conventional deposition processes.

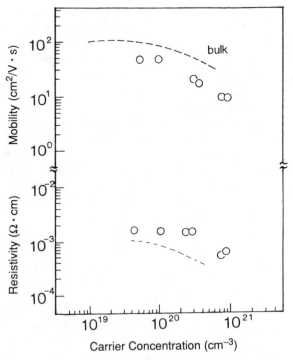

Figure 50. Dependence of the electron mobility and film resistivity on carrier concentration for epitaxial Si films grown on Si(100) substrates by Hg-sensitized photodeposition [after Jia et al. (*271*), reprinted by permission].

Given a sufficiently short wavelength optical source, monosilane itself can, of course, also be photodissociated efficiently. A recent example is the work of Milne et al. (*272*) who built a hydrogen discharge lamp directly into the photochemical reactor. By photodissociating silane with the 160 nm (Werner band) radiation from the lamp, amorphous hydrogenated Si films were grown at 250°C, a total pressure of ~ 0.8 Torr and a deposition rate of $4 \, \text{Å} \cdot \text{s}^{-1}$. The optical characteristics of these films (see Table 14) are comparable to or superior to those for material deposited by laser or RF plasma. The photoconductivity ratio (i.e., AM1 conductivity to dark conductivity) is 6–7 orders of magnitude and the optical bandgap was measured to be 1.75 eV. Despite these excellent results, the deep VUV ($\lambda \lesssim 160$ nm) wavelengths required for the efficient decomposition of SiH_4 have resulted in the preponderance of direct photodecomposition studies being carried out with Si_2H_6 or Si_3H_8 as the precursor. Early attempts at photodissociating Si_2H_6 with Hg lamp radiation resulted in deposition rates that were low compared to those obtained by Hg photosensitization. Mishima et al. (273–275) observed deposition rates for amorphous Si films as high as $15\,\text{Å}\cdot\text{min}^{-1}$ when Si_2H_6 was directly photodissociated with the 185 and 254 nm lines from a mercury lamp. The lamp intensity was $\sim 80\,\text{mW}\cdot\text{cm}^{-2}$, the quartz or Si substrate temperature was varied between 200 and 400°C, and the gas feedstock mixture consisted of 1% Si_2H_6 in He, with N_2 acting as the carrier gas. In the 200–400°C temperature range investigated, the photochemical film growth rate was found to be virtually independent of temperature, a result consistent with the activation energies (101, 270) that were mentioned earlier. An investigation (276) of the early stages of the photo-CVD deposition of Si by the direct photodissociation of Si_2H_6 with an Hg lamp showed that the concentration of SiH_3 species in the film decreases with the deposition time.

Fuyuki et al. (277) also explored the use of a D_2 lamp in depositing a-Si:H films in the 200–300°C temperature range. No change in the $15\,\text{Å}\cdot\text{min}^{-1}$ deposition rate was observed in this temperature interval but substituting a microwave-excited Xe lamp for the D_2 source improved the deposition rate. For ~ 120 and 50 W of input power to the lamp, the a-Si:H film deposition rate was observed to be 75 and $10\,\text{Å}\cdot\text{min}^{-1}$, respectively. By changing the optical window on the reactor from CaF_2 or MgF_2 to quartz, it was shown that film growth requires radiation wavelengths below 180 nm. Similar results were obtained by Kawasaki et al. (278), who used a pulsed Xe arc lamp, operating at a repetition frequency of 50 Hz, to photodissociate Si_2H_6. Only photons of wavelength $\lambda < 200$ nm were found to be effective for photodeposition, and the activation energy for the process in the 150–250°C range was measured to be 0.15 eV.

At higher substrate temperatures, increased deposition rates and the growth of epitaxial films were reported by Gonohe and co-workers (279), who illuminated a Si_2H_6/N_2 gas mixture and Si(100) substrates with the VUV radiation from a D_2 lamp. Epitaxial films were obtained at 650°C, and at that substrate temperature the deposition rate was measured to be $\sim 900\,\text{Å}\cdot\text{min}^{-1}$. Over the 550–650°C temperature range, the activation energy for the deposition process is 23 kcal/mol $\simeq 1$ eV, which is roughly 66% of the value for thermal deposition (conventional CVD). Consequently, at the low end of the temperature region studied (550°C), the photo-CVD deposition rate was found to be more than double the thermal rate. The assistance of the incident VUV photons in cleaning the Si surface was also clearly demonstrated. Epitaxial films were also grown by Yamazaki et al. (280) on 4-in. Si(100) wafers in the reactor illustrated in Figure 51. The susceptor was heated by a bank of IR lamps, and the UV lamps produced a fluence of 1.2 W/cm² ($\lambda < 300$ nm) at the substrate. Provision was made for doping the growing Si films either n- or p-type, and p^+–n^+ junctions were grown continuously by switching the dopant precursor molecule in the gas stream from B_2H_6 to PH_3. Figures 52

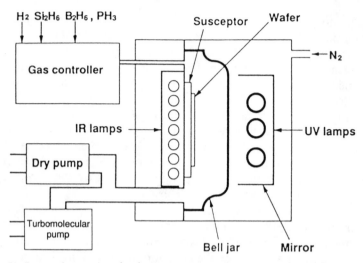

Figure 51. Reactor for growing n^+-p^+ epitaxial layers on 4-in.-diameter Si wafers by photo-CVD [after Yamazaki et al. (*280*), reprinted by permission].

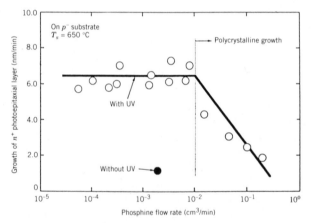

Figure 52. Dependence of growth rate on the PH_3 flow rate for an n^+ layer on a p-type Si substrate (*280*). The film was grown at a substrate temperature of 650°C, and the growth rate measured in the absence of UV illumination of the substrate is also shown.

and 53 show the growth rate and carrier concentrations, respectively, as a function of the PH_3 flow rate for an n^+ layer grown on a p-type substrate at 650°C. For low mass flow rates the film growth rate is essentially constant at $\sim 65\,\text{Å·min}^{-1}$, but above $\sim 10^{-2}$ sccm epitaxial films can no longer be grown. Abrupt p^+-n^+ layers can be grown continuously in this manner, and Figure

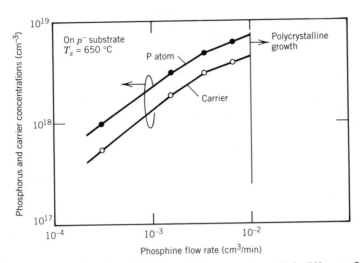

Figure 53. Variation of the phosphorus and carrier concentrations with the PH_3 mass flow rate for an n^+ layer photodeposited at 650°C (ref. *280*).

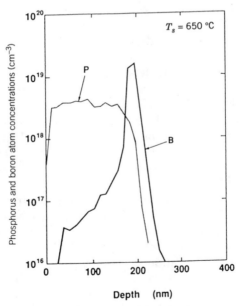

Figure 54. Phosphorus and boron depth profiles for the p^+-n^+ structure grown in ref. *280* by photo-CVD (reprinted by permission).

54 shows the phosphorus and boron concentration profiles that were obtained by SIMS (secondary ion mass spectrometry) analysis.

Photodecomposition of Si_3H_8 with a deuterium lamp has been reported by Fujiki et al. (281), who deposited polycrystalline films with preferred $\langle 110 \rangle$ or $\langle 100 \rangle$ orientation and thicknesses ranging from 2.5 to 10 μm. After photolyzing Si_2H_6 and Si_3H_8 with an Hg lamp producing 3.5 W/cm^2 at 185 nm, Kumata et al. (282) found that the deposition rate for a-Si:H films grown from Si_3H_8 was five times that for Si_2H_6 as a precursor. The conductivity of films deposited from Si_3H_8 at 300°C [10^{-5} ($\Omega \cdot$cm)$^{-1}$] makes these films well suited for incorporation into solar cells. Recently, Kim and co-workers (283) prepared p–i–n a-Si:H solar cells from films photodeposited from Si_2H_6 and measured efficiencies $>11\%$.

5.1.2.3. Laser Photodeposition

Amorphous, polycrystalline, and epitaxial Si films have been deposited from several different precursors with lasers and in both parallel and perpendicular configurations. Andreatta et al. (193) deposited polycrystalline Si films on quartz substrates at rates up to 9.8 Å·s^{-1} with a weakly focused KrF laser and peak intensities of ~ 20 MW·cm^{-2}. The high intensities required are a consequence of the dominance of two-photon dissociation of monosilane at 193 nm (ref. 149). Film thicknesses up to ~ 2000 Å were obtained with $\gtrsim 400$-s exposure times and 10% SiH_4 in N_2 gas mixtures (total pressure of several hundred Torr). Films were also deposited with $Si(CH_3)_4$ as the precursor.

Several groups (284–286) have reported the deposition of hydrogenated amorphous films from the photodissociation of Si_2H_6 and Si_3H_8 with an ArF excimer laser. Deposition rates ranging from 100 to 300 Å·min^{-1} were recorded, and Table 14 (175) provides a summary of the properties of these films, including their photoconductivity (at AM1), σ_{ph}, and dark conductivity, σ_D, that are critical parameters in determining the suitability of the films for photovoltaic applications. For the sake of comparison, Table 14 also gives similar data for a-Si:H films photodeposited with lamp sources [Hg (direct photodissociation) and windowless He$^+$ and H_2].

An example of the performance of these films in electronic devices is provided by the thin film transistor (TFT) fabricated by Hiura et al. (287). ArF laser grown a-Si:H films having electrical properties similar to those of Table 14 ($\sigma_D = 2.6 \times 10^{-10}$ ($\Omega \cdot$cm)$^{-1}$; $\sigma_{ph}/\sigma_D \sim 10^3$) was incorporated into the TFT structure shown schematically in Figure 55, and Figure 56 compares the drain current–gate voltage characteristics for TFTs fabricated from plasma CVD and laser-deposited a-Si:H films.

Epitaxial Si films were deposited by excimer laser-assisted CVD by Yamada et al. (288) in the 600–650°C temperature range. Not only did the presence

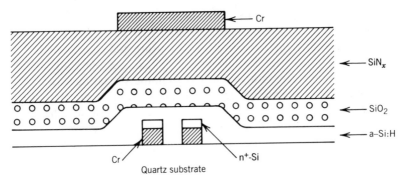

Figure 55. Thin film transistor fabricated with photo-CVD a-Si:H and SiO$_2$ films [Hiura et al. (287), reprinted by permission].

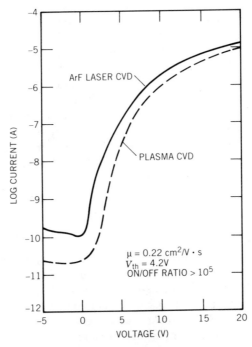

Figure 56. Drain current–gate voltage characteristic curve for the thin film transistor of Figure 55.

of UV photons at the substrate triple the film growth rate, but crystallinity and hole mobilities were both markedly improved with the UV laser beam and particularly at the lowest substrate temperature studied (600°C). Lian and co-workers (289) have successfully grown epitaxial Si films on Si(100) at considerably lower temperatures (330°C) by photodissociating between 3 and 20 Torr of Si_2H_6 at 193 nm in parallel geometry. Both *ex situ* (HF dip) and *in situ* cleaning of the substrates were employed to obtain films having low defect (stacking fault and dislocation loop) densities: $< 10^6 \, cm^{-2}$. The substrate temperature restricts the deposition rate to below $4 \, Å \cdot min^{-1}$, and film growth was attributed primarily to the gas phase generation and subsequent adsorption of silylene, SiH_2. At lower Si_2H_6 pressures (0.01–0.10 Torr), thin Si films have also been deposited on quartz (290) or sapphire (291) substrates. In a 1% Si_2H_6/He gas mixture, Tanaka, Deguchi and Hirose (290) demonstrated spatially selective deposition of Si onto quartz by directing the 193-nm laser beam through a metal mask (21.5- or 130-μm stripe pattern) and onto the substrate. The dominant mechanism in film deposition was determined to be dissociative adsorption of Si_2H_6 onto the substrate in the presence of the optical field.

Before leaving this section, the work of Frieser (292) and Ishitani et al. (293) in UV-enhanced epitaxy should be noted. Frieser reported that the deposition of epitaxial Si films from the pyrolysis of Si_2Cl_6/H_2 gas mixtures was accelerated when the substrate was irradiated by a mercury arc lamp. Since the lamp's 5 eV ($\lambda \sim 254$ nm) photons are not sufficiently energetic to rupture the Cl_3Si—$SiCl_3$ bond, the observed enhancement was attributed to surface interactions. Similar results were obtained by Ishitani and co-workers (293), who measured significant increases in the deposition rate of epitaxial Si (by CVD from SiH_2Cl_2, dichlorosilane) when Hg–Xe UV radiation was directed onto the substrate. Specifically, the growth rate at 900°C rose from 1.8 to 2.8 μm/min, which was attributed to the photodissociation of the $SiCl_2$ radical at the substrate surface.

5.1.3. Germanium

Fewer studies have been concerned with the photodeposition of elemental germanium. Polycrystalline films up to 1500 Å in thickness were deposited by Andreatta et al. (193) by photodissociating germane (GeH_4) with 193- or 248-nm excimer radiation. The perpendicular geometry experiments yielded deposition rates of 4–$5 \, Å \cdot s^{-1}$ for ArF laser intensities of $2 \, MW \cdot cm^{-2}$. Higher 248-nm intensities ($\sim 8 \, MW \cdot cm^{-2}$) were required to reach the maximum film deposition rate ($\sim 6 \, Å \cdot s^{-1}$) for that source wavelength. No film growth was observed when the laser wavelength was 308 nm (XeCl laser), which demonstrates the photochemical nature of the process. X-ray measurements showed the films to consist of randomly oriented grains up to 0.5 μm in diameter.

Epitaxial Ge films were grown by Eden et al. (294) on NaCl(100) substrates at substrate temperatures as low as 120°C. As mentioned previously, the activation energy for the deposition process was determined to be 85 ± 20 meV (101, 294). Polycrystalline films were deposited on quartz substrates ($T_S \leqslant 250$°C) with grain sizes ranging from 0.3 to 0.7 μm.

In parallel geometry, thin epitaxial films have been deposited on GaAs(100) substrates at temperatures as low as 285°C by photodissociating GeH_4 with an ArF laser (295, 296). The position of the laser beam was maintained at $\geqslant 2$ mm above the substrate, and the dominant role played by the laser appears to be the conversion of GeH_4 to Ge_2H_6, which migrates > 10 mean free paths to the substrate, where it subsequently pyrolyzes. Etching the substrate in situ by flowing 5% HCl (in He) through the reactor for 5 min was found to be necessary for obtaining epitaxial films. After ~ 400–700 Å of epitaxial growth, the growing film switches abruptly to amorphous material, which is apparently due to the buildup of carbon (from the vacuum system) at the solid/vapor interface.

More than an order of magnitude increase in the Ge film growth rate is

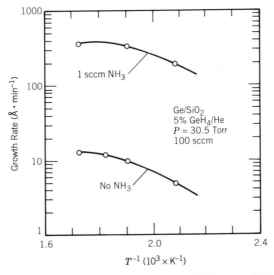

Figure 57. Enhancement in the growth rate of photo-CVD Ge films when NH_3 is added to the He/5% GeH_4 gas stream. The substrate temperature is varied from 205 to 305°C [after Kiely et al. (140)].

observed if small amounts of ammonia are added to the GeH_4/He gas flow stream (see Figure 57). Kiely and co-workers (140) demonstrated that, despite the acceleration in the deposition rate, the Ge films are nevertheless epitaxial. Upon absorbing a 193-nm photon, NH_3 produces an H atom which subsequently decomposes GeH_4 by hydrogen abstraction [see Eqn. (26) in Section 2.2.3.1]. Furthermore, the absorption cross section of ammonia at 193 nm is almost 3 orders of magnitude larger than that for GeH_4 at the same wavelength ($\sigma_{NH_3} \sim 10^{-17}$ cm^{-2}; σ_{GeH_4} 2–3 × 10^{-20} cm^2). In these experiments, then, NH_3 acts as a photosensitizer without incorporating into the Ge crystalline film to any significant degree (140).

Structures consisting of alternating layers of amorphous Ge and Si were deposited entirely by photochemical processes by Lowndes et al. (48). Film deposition was carried out in parallel geometry (laser beam-to-substrate distance $\geq$ 1 mm) by photodissociating Si_2H_6 or GeH_4 in the reactor. Ge and Si layer thicknesses of 5.4 ± 0.2 and 10.7 ± 0.4 nm, respectively, were deposited in fabricating a nine-layer structure. Germanium/silicon alloy films have also been photodeposited from GeH_4/Si_2H_6 mixtures using an ArF laser (297). For substrate temperatures below 350°C, deposition occurred entirely by photochemical processes and deposition rates ranged from 150 to 300 Å·min^{-1}. The ratio of Si to Ge concentrations in the films was found to be a factor of 3 higher than the Si_2H_6:GeH_4 partial pressure ratio, which suggests the importance of cross reactions between Ge- and Si-containing species in the film deposition process. Alternative precursors for the photo-CVD of Ge were explored by Stanley et al. (298), who deposited polycrystalline Ge films from ethylgermane and diethylgermanium. Although an IR (CO_2), rather than UV, laser was used, the photochemical nature of the deposition process was suggested by the requirement that the laser be tuned to an IR absorption band of the precursor for deposition to occur.

5.1.4.   Summary

Amorphous, polycrystalline, and epitaxial Si and Ge have now been grown by photo-CVD with several precursors and in a variety of reactor configurations. Film quality, as judged by electrical, chemical, and structural characteristics, is generally comparable to those for films deposited by conventional techniques such as plasma CVD. Amorphous, hydrogenated (a-Si:H) films deposited by laser or windowless lamp, for example, exhibit dark and AM1 photoconductivities that are typically $\lesssim 10^{-9}(\Omega\cdot cm)^{-1}$ and $> 10^{-4}(\Omega\cdot cm)^{-1}$, respectively, which matches or exceeds the characteristics of films deposited by other methods. Epitaxial Si films have been grown at substrate temperatures below 200°C, and the reduced growth temperatures have permitted p^+-n^+ layers to be grown with unusually abrupt doping profiles.

5.2. II–VI BINARY AND TERNARY COMPOUNDS

Photo-assisted CVD is a particularly attractive deposition technique for those devices and materials, such as the II–VI compound semiconductors, that are sensitive to processing temperature. As Irvine (299) has pointed out, interdiffusion in the II–VI ternary compound $Cd_xHg_{1-x}Te$, for example, is typically $\geqslant 100 \, \mu m$ at a growth temperature of 600°C but is $< 1 \, \mu m$ for temperatures at or below 400°C. Furthermore, the electrical properties of the II–VI materials are difficult to control, and lowering the growth temperature will certainly minimize, for example, dislocation densities and ameliorate the well-known self-compensation effect in II–VI crystalline layers. Efforts to fabricate multilayer structures with abrupt interfaces and well-behaved electrical characteristics (by substitutional doping) in the II–VI compounds, therefore, hinge on an ability to deposit high-quality films at reduced temperatures. While the temperatures at which the binary II–VI semiconductors ZnSe, ZnS, CdS, CdSe, CdTe, and HgTe are deposited by MOCVD can be reduced to below 250°C by resorting to less thermally stable precursors, photo-assisted deposition of these compounds at temperatures as low as 150–200°C has yielded epitaxial films of high quality. Such results are especially interesting and timely in view of the recent demonstration of blue-green semiconductor lasers based on the fabrication of p–n junctions in II–VI materials (ZnSSe). This section will review the remarkable progress that has been realized over the past several years in the photo-CVD growth of epitaxial II–VI layers and modulated structures.

5.2.1.   Zinc Compounds

The earliest work in II–VI compound photodeposition centered on the zinc-containing binary compounds ZnSe and ZnO. Johnson and Schlie (225) were the first to demonstrate the deposition of zinc selenide films by irradiating mixtures of $Zn(CH_3)_2$ and $Se(CH_3)_2$ vapor with an Hg arc lamp. Although the films were not stoichiometric, film thicknesses up to $\sim 0.6 \, \mu m$ over more than $3 \, cm^2$ of surface were obtained. Exposure periods of 1–2 h were necessary to deposit these films owing to the low fluence of the Hg lamp for wavelengths below $\sim 245 \, nm$, which was found to be the wavelength threshold for the photodissociation process. Polycrystalline zinc monoxide films were subsequently reported by Solanki and Collins (178), who employed an ArF or KrF excimer laser to photodissociate $Zn(CH_3)_2$ and NO_2 (or N_2O). Both parallel and perpendicular geometries were explored, but the highest optical quality (defined with regard to transmission) and lowest resistivity films were obtained by irradiating the silicon substrate. Deposition rates up to $18 \, \mu m \cdot h^{-1}$ were measured and the uniformity in film thickness over $10 \, cm^2$ of surface area was $\pm 5\%$. Shimizu et al. (*164*) were able to grow epitaxial ZnO($11\bar{2}0$) films on ($01\bar{1}2$) sapphire at 500°C and epitaxial (0001) layers on (0001) sapphire at 450°C. A KrF laser was again the optical source, the reactant gases were NO_2 and diethylzinc, and direct irradiation of the surface was required in order to obtain crystalline films. Surface irradiation also improved the surface morphology and electrical characteristics of the ZnO layers, which was attributed to photoexcitation of adsorbed species and photoenhancement of surface mobilities. When the laser illuminated the substrate, ZnO films having higher mobilities (4–10 cm^2/V·s) and carrier concentrations (5–8 × 10^{18} cm^{-3}) and lower resistivities (0.1–2.0 Ω·cm) than those for films grown in the absence of direct irradiation were obtained.

With a mercury lamp (low pressure) as the optical source, c-axis-oriented polycrystalline ZnO films have been grown for substrate temperatures above 200°C (*300*). The precursors were $Zn(CH_3)_2$ and either CO_2 or O_2, and the film deposition rate was observed to be closely tied to the reactor geometry. For ZnO films grown at 300°C in a vertical configuration reactor, the deposition rate for photodeposited films was roughly a factor of 2 greater than that for thermally grown films.

In 1985, Ando et al. (*234*) reported the photodeposition of ZnSe epitaxial films at temperatures as low as 150°C. Diethylzinc and dimethylselenide were the alkyl precursors, and the UV radiation was provided by a low-pressure Hg lamp. At 400°C, the presence of the UV improved the film deposition rate by a factor of 2. The impact of UV photons on the ZnSe film growth rate is illustrated in Figure 58. For substrate temperatures below ~ 350°C, deposition occurs only through photo-assisted processes. Epitaxial films were obtained by photo-assisted MBE at 450°C, and between 150 and 350°C the ZnSe films were polycrystalline. Irradiation of ZnSe polycrystalline films during growth on glass substrates produced a preferential (111) orientation. Recently, Fujita and co-workers (*301–303, 427*) have enhanced the rate of growth of epitaxial ZnSe (*301–303, 427*) and ZnS (*303*) films by illuminating the substrate with Xe arc lamp radiation having a maximum intensity at the substrate of 100 mW·cm^{-2}. Experiments (*301, 302*) carried out in an MOCVD reactor at 200-Torr total pressure with dimethylzinc and dimethylselenium precursors showed that, in the absence of the external UV radiation, ZnSe film deposition was $< 0.5\,\mu$m·h^{-1} for substrate temperatures below 400°C. Once the UV beam is directed onto the surface, however, deposition rates as high as 1.5 μm·h^{-1} (250 Å·min^{-1}) were observed at temperatures down to 300°C. For a substrate temperature of 350°C, the ZnSe deposition rate was 1.5 μm·h^{-1} for 47 mW·cm^{-2} of lamp intensity. In the absence of lamp radiation, the deposition rates fell to $< 0.04\,\mu$m·h^{-1}. It was confirmed that the gas phase absorption of the source radiation was negligible and the primary role played by the lamp was to excite surface reactions. Fujita et al. (*303*) later showed that the longest wavelengths at which the deposition

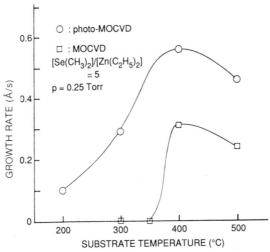

Figure 58. Growth rates for ZnSe films on GaAs(100) as a function of temperature for photo-assisted and conventional MOCVD deposition [reprinted from Ando et al. (234) by permission].

rate enhancement occurs are ~ 500 and 335–350 nm for ZnSe and ZnS, respectively, which strongly suggests that carriers photogenerated at the growing surface are responsible for the observed effect. At 400°C, zinc sulfide films are deposited at rates exceeding $3 \, \mu m \cdot h^{-1}$ when the GaAs(100) substrate is irradiated but is an order of magnitude lower without the lamp. Figure 59 shows the data that were reported by Fujita et al. (303) over the 280–600°C temperature range. Fujita and co-workers (427) also demonstrated that irradiating the surface with photons in the $400 \leqslant \lambda \leqslant 800$ nm region does not improve the growth rate but does enhance the incorporation of donor impurities into the crystalline ZnSe film. To obtain the highest quality epilayers, they restricted the photon wavelength to $\lambda < 400$ nm.

Similar studies were carried out for ZnSe and $ZnS_x Se_{1-x}$ in which epilayers were grown by molecular beam epitaxy while the substrate was illuminated by He–Cd laser (325, 442 nm) or Hg–Xe lamp radiation (304–306, 428). Surface irradiation not only enhanced film growth rates and surface morphology but also significantly improved the material's optical quality, as evidenced by increased free exciton emission intensities. The latter was observed for optical intensities at the substrate as low as $1 \, mW \cdot cm^{-2}$ and apparently stems from a decline in the nonradiative recombination center density. ZnSe epitaxial layers were grown at temperatures as low as 150°C. A common theme in these studies was that e–h pairs photogenerated at the surface and the interaction of photons with physisorbed atoms, apparently resulting in a reduction of the activation barrier for desorption, played major roles in realizing low temperature film growth. In the same vein, Ogawa et al. (307) grew homoepitaxial

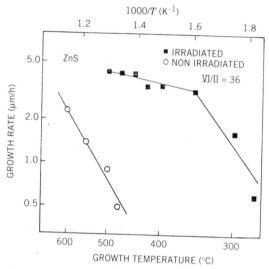

Figure 59. Dependence of the ZnS growth rate on temperature showing the influence of substrate irradiation [after Fujita et al. (*303*)].

ZnTe films by MOCVD from DMZ and diethyltelluride at temperatures as low as 350°C with Ar$^+$ laser (488 nm) irradiation of the (100)-oriented substrate (laser fluence typically 40 mW·cm^{-2}).

Polycrystalline Mn-doped ZnS (ZnS:Mn) films photodeposited under high-pressure Hg lamp irradiation were examined for their potential in thin film electroluminescent devices (*308*). Although the photo-assisted films were grown at temperatures 50–100°C lower than standard MOCVD films and the grain sizes were larger, electroluminescent devices fabricated from the photo-CVD films exhibited comparatively low luminance due to sulfur vacancies.

Epitaxial ZnS and ZnSe films have also been deposited on GaAs with excimer laser radiation (193 and 248 nm) by Fujita et al. (*38*) and Shinn and co-workers (*309*), respectively. In both cases, the UV laser beam passed parallel to the substrate and epitaxial films were obtained in the 100–150°C range for ZnS and at temperatures as low as 200°C for ZnSe. Although single-pulse laser energies were varied from 7.5 to 50 mJ, deposition rates did not exceed 1 μm·h^{-1}.

5.2.2. Cadmium and Mercury Binary and Ternary Compounds

Considerably more effort has been devoted to the Cd- and Hg-based II–VI films. Virtually all of the photodeposition studies reported to date have relied upon the alkyls as sources for one or both of the Column IIB and VIB atoms.

HgTe. The first photodeposited epitaxial II–VI film, HgTe, was grown by Irvine and coworkers (310, 311) from diethyltelluride, $Te(C_2H_5)_2$, and mercury vapor by Hg photosensitization. Films $\sim 1\,\mu m$ thick were deposited at rates up to $\sim 2\,\mu m \cdot h^{-1}$ in the 240–310°C interval, and epitaxial films could be grown at temperatures as low as ~ 180°C (310, 312). In the 180–250°C range, no film deposition occurs in the absence of UV radiation (313). Increasing the precursor partial pressures and/or Hg lamp intensity will offset the decline in film growth rate at low temperatures, and it has been estimated that a deposition rate of $1\,\mu m \cdot h^{-1}$ could be obtained at 180°C for UV intensities at the substrate of $100\,W \cdot cm^{-2}$ (313). Utilizing the radiation from an Hg–Xe arc lamp, Ahlgren et al. (173, 314) also photochemically deposited HgTe films at temperatures as low as 182°C and successfully fabricated HgTe–CdTe superlattices (> 50 periods) by alternately introducing $Te(C_2H_5)_2$ and $Cd(CH_3)_2$ into the reactor (in the presence of Hg vapor). HgTe film deposition rates were $< 0.1\,\mu m \cdot h^{-1}$ at 182°C when diethyltellurium was the Te precursor, but at 240°C growth rates rose to roughly $1\,\mu m \cdot h^{-1}$ when diisopropyl was used as the precursor. Laser photochemical vapor deposition of HgTe was recently reported by Fujita, Fujii and Iuchi (38). With KrF or ArF lasers and a parallel configuration, epitaxial films were grown on CdTe(111) substrates at deposition rates up to $\sim 1\,\mu m \cdot h^{-1}$. Early attempts to pattern HgTe on CdTe at 300°C by a projection arrangement using a frequency-doubled Ar^+ laser (257 nm) have been successful (129, 315).

CdTe. Of the II–VI materials, the highest quality photodeposited films have been those for the binary CdTe. Kisker and Feldman (316, 317) grew epitaxial CdTe films on GaAs(100) at substrate temperatures down to 250°C with a low-pressure Hg lamp. For dimethylcadmium and diethyltelluride as the precursors, obtaining epitaxial films required that the $Cd(CH_3)_2:Te(C_2H_5)_2$ partial pressure ratio be 1.5:1.0. Long-term film growth rates up to $2\,\mu m \cdot h^{-1}$ were recorded at 350°C. Excimer laser-assisted photoepitaxy of CdTe was demonstrated by Zinck et al. (37, 179). Growth rates as high as $2\,\mu m \cdot h^{-1}$ were recorded at a deposition temperature of 165°C, and the films, deposited on GaAs(100) substrates, were oriented (111). The precursors for these studies were also $Cd(CH_3)_2$ and $Te(C_2H_5)_2$, but partial pressures for both reactants were low: 10^{-3}–10^{-2} Torr. Also, while both 193- and 248-nm radiation was investigated, most of the effort focused on the KrF laser. An especially interesting aspect of the results (179) was the observation that the highest quality epilayers were obtained with parallel optical geometry (where the laser does not irradiate the substrate). For these films, carbon and oxygen concentrations were below the limits of detectability by Auger analysis. When the substrate was illuminated by the excimer laser, on the other hand, the film quality deteriorated noticeably.

Cody, Sudarsan and Solanki (318, 319) successfully grew epitaxial CdTe films on GaAs with an Hg lamp. For only $100\,mW \cdot cm^{-2}$ of radiation incident on the surface ($200 \leqslant \lambda \leqslant 250\,nm$), maximum deposition rates were found to be as high as 13 and $9\,\mu m \cdot h^{-1}$ for CdTe and GaAs(100) substrates, respectively (318). For $(CH_3)_2Cd$ and $(C_2H_5)_2Te$ partial pressures of 0.4 Torr and 1.2 Torr, respectively, photodeposition rates were typically $6\,\mu m \cdot h^{-1}$ at 250°C (319).

With a frequency-doubled Ar^+ laser as the optical source, Irvine et al. (315) demonstrated that, in the presence of the 257-nm photons, the growth of CdTe on GaAs[(100)2° →(110)] at 300°C was enhanced by factors ranging from 4 to 60, depending on which of three tellurium precursors was used. Bar and mesa structures were photopatterned onto GaAs substrates by this process, and feature sizes down to $50\,\mu m$ were obtained. SIMS analysis of the films showed the CdTe to be of high purity, but impurities were concentrated at the CdTe/GaAs interface. It appears that the adsorption of the Te-containing species (or Te itself) and the desorption of CH_3 and C_2H_5 groups are the surface processes that control the growth process and permit spatially selective epitaxy to occur.

The growth of epitaxial CdTe at lower temperatures (165°C) and in *parallel* geometry was modeled recently by Liu and co-workers at UCLA and Hughes Research Laboratories (320). The simulations indicate that, for an optical source wavelength of 248 nm, Te is adsorbed onto the substrate and the growth rate is limited by the transport of Cd atoms to the surface. That is, the CdTe film growth rate is controlled by the rate at which alkyl precursors (and the Cd-alkyl, in particular) are photodissociated by the optical source. Consequently, in this case, the growth-rate-limiting mechanisms are gas phase rather than surface reactions. Excess Te on the surface is removed by methyl radicals, and those CH_3 species not required for Te removal are expected to recombine on the surface and desorb as ethane. The requirement that unnecessary Te atoms be removed by CH_3 radicals imposes an upper limit on the Te-alkyl partial pressure. Increasing the Te precursor pressure beyond this maximum results in the co-deposition of elemental Te with CdTe.

The model predicts film growth rates that are within 10% of the measured values. Also, the calculated and measured growth rates are linear with the dimethylcadmium partial pressure and laser power, which confirms the limitations imposed by the rate of gas phase photolysis of the Cd-alkyl. One concludes that large-area epitaxial CdTe films of high purity can be grown on GaAs at temperatures below 200°C if the Cd- and Te-alkyl partial pressures and mass flow rates are adjusted properly and the gaseous region illuminated by the UV optical source is immediately adjacent to the substrate. In order to better understand the growth of binary CdTe or HgTe films by photo-assisted processes, it is necessary that, as mentioned in Ref. 320, the distribution of products produced when Te alkyls are photodissociated should be measured. Also, for surface-illuminated growth, more detailed studies of photon-driven reactions at 257 and 248 nm are needed.

Cd_xHg_{1-x}. The ternary alloy $Cd_xHg_{1-x}Te$ was deposited by Irvine et al. (313, 321, 322), who grew thin epitaxial films by Hg photosensitization. With diethyltelluride, dimethylcadmium, and Hg vapor as the precursors and a 3-kW Hg arc lamp, epitaxial films 1.3 μm thick were grown at temperatures down to 250°C. The use of He rather than H_2 as the carrier gas was found to be necessary for the growth of epitaxial films. Morris (323) subsequently grew epitaxial $Cd_{0.8}Hg_{0.2}Te$ films with an ArF laser in parallel geometry and, rather than Hg vapor, employed $Hg(CH_3)_2$ (dimethylmercury) as the Hg donor molecule. With CdTe substrates, growth rates of $\sim 4\,\mu m \cdot h^{-1}$ and >2-μm film thicknesses were obtained at a temperature of 150°C. Since the laser did not irradiate the substrate, transient heating of the growing film by the optical source was avoided. Recently, Cody et al. (319) obtained similar growth rates ($4\,\mu m \cdot h^{-1}$) for 20% CdTe/80% HgTe alloy films that were deposited at 250°C with the aid of a 2-kW Hg lamp. The best film quality resulted from using methylallyltelluride as the Te precursor and the variation in film composition, x (i.e., $Cd_xHg_{1-x}Te$), was only 4% over 1 cm^2 of surface area.

In summary, the II–VI compound semiconductor films have benefited significantly from the introduction of photon-assisted growth processes. In the gas phase, the production rates of methyl radicals, for instance, and metal atoms can be controlled for optimal growth rates ($1-10\,\mu m \cdot h^{-1}$) and minimal impurity levels. Surface illumination clearly promotes the decomposition of precursors and stimulates the desorption of products. The already-demonstrated growth of high-quality films at temperatures below 250°C (and, in several cases, below 200°C) indicates that the fabrication of compositionally modulated structures by photo-CVD is an attractive route for growing these temperature-sensitive materials.

5.3. III–V COMPOUND FILMS

Interest in the growth and doping of III–V compound films by photo-assisted processes is a natural consequence of the importance of these materials to multilayer optoelectronic devices. As was the case with II–VI films, interdiffusion (and impurity redistribution, in particular) in compositionally and doping-modulated structures and thermal stress are critical issues driving a wide-ranging research effort in low-temperature processing. Although the first efforts to photodeposit III–V films were reported early in the last decade, the application of photo-CVD to this family of materials initially developed more slowly than did metal or II–VI film deposition but progress has recently accelerated significantly. Because of surface diffusion rates, the optical source has most often *assisted* the growth of III–V films by conventional (thermal) techniques such as MOCVD or MBE. (An exception to this generalization is GaN, which has recently been grown by photo-CVD at temperatures (600°C) where thermal growth is negligible.) In contrast, epitaxial II–VI films have (as discussed in the last section) been grown at temperatures as low as 150°C, where measurable deposition will not occur in the absence of external UV radiation. The presence of an optical beam at the substrate during the growth of III–V films has been demonstrated to yield several beneficial results such as improvements in surface morphology, modification of carrier concentrations, and—in several cases—increased deposition rates. Moreover, doping of GaAs can now be controlled optically and doping superlattices have been fabricated with an excimer laser. The following subsections will provide an overview of the work reported to date in this area, with emphasis on the photo-assisted deposition of GaAs.

5.3.1. GaAs

Most of the effort in photodepositing III–V films has concentrated on the binary compound GaAs. Pütz et al. (*325*) examined the growth of GaAs in the 510–650°C range and found that the presence of UV photons from an Hg lamp improved the maximum (saturated) growth rate from 5 to 8 μm·h^{-1} at 510°C. Also, they observed that "in all epitaxial runs improved surfaces were obtained upon illumination" (*325*) and that mirror-like surfaces were produced even at temperatures as low as 510°C. Similar results were obtained subsequently by Balk and co-workers (*326*), who also employed ArF and XeF excimer lasers (irradiating the substrate at normal incidence) in addition to the Hg lamp. While the film growth rate was observed to double for 193-nm illumination of the substrate, no effect was detected for $\lambda = 351$ nm, which points to the photochemical nature of the effect as opposed to transient heating of the substrate or the dominance of photo-generated carriers. Figures 60 and 61 illustrate the improvement in film deposition rates that were obtained in (*326*) by irradiating the substrate with low fluences (< 1 mW·cm^2) of continuous 185-nm radiation or the pulsed beam from an excimer laser. The comparative impact of Hg lamp radiation on the growth of GaAs from one of three Ga-alkyls—trimethylgallium (TMG), triethylgallium (TEG), and triisobutylgallium (TIBG)—is shown in Figure 60. For each of the Ga precursors studied, sets of data (growth rates measured with and without surface photons) are given for two substrate temperatures. One temperature (the higher of the two) lies in the region where the reactor kinetics are controlled by the transport of reactants to the substrate, whereas at the second (lower) temperature, the film growth rate is limited by surface reactions. As would be

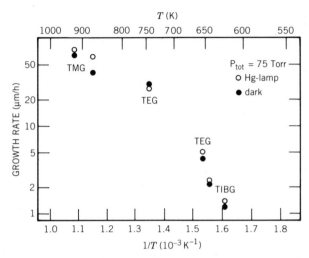

Figure 60. Growth rate data for the photo-CVD of GaAs with an Hg lamp [reprinted by permission from Balk et al. (*326*)]. The Ga precursors studied were trimethylgallium (TMG), triethylgallium (TEG), and triisobutylgallium (TIBG). For each precursor, growth rates were measured at two temperatures. One data set was acquired at a temperature (the higher of the two) where the reactor kinetics are limited by the transport of reactants to the surface, and the other at a lower temperature where surface reactions limit the growth rate. Note that the percentage improvement in growth rate is larger for the more stable Ga-alkyl precursors (i.e., TIBG and TEG). The experimental parameters were as follows: TMG data—$P_{TMG} = 0.32$ Torr, $P_{AsH_3} = 1.7$ Torr; TEG data—$P_{TEG} = 0.19$ Torr, $P_{AsH_3} = 1.3$ Torr; TIBG data—$P_{TIBG} = 22$ m-Torr, $P_{AsH_3} = 128$ mTorr. The total reactor pressure in all cases is 75 Torr.

expected, the enhancement in film growth rate is largest at the lower of the two temperatures studied for each of the three precursors. Only in this region will the beneficial effects of surface irradiation, including the photodissociation of adsorbed Ga alkyls (or their fragments such as *mono*methylgallium) and photo-stimulated desorption, become apparent. Also, the percentage increase in growth rate is more pronounced for the more thermally stable precursors. Said another way, the percentage enhancement in growth rate follows the sequence: TMG → TEG → TIBG. This result clearly demonstrates the benefit of resorting to photoprocesses to decompose one or more precursors rather than relying on temperature alone. In the former case, dismantling the molecule can be separated from the surface temperature and it is now possible to optimize separately: (1) the number density of reactants (or their fragments) at the surface, and (2) the surface mobilities of adsorbed species. Several considerations may motivate one to resort to more thermally stable precursors. The control afforded by optically induced growth is one issue; another is that more thermally stable precursors are often less toxic. In any case,

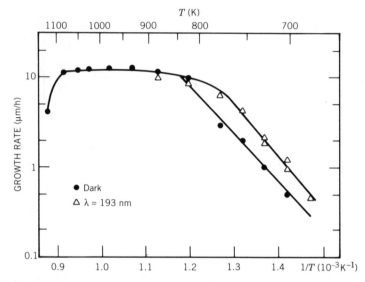

Figure 61. Dependence of the photo-assisted GaAs deposition rate on temperature with and without 193-nm (ArF) excimer laser radiation at the substrate [reprinted by permission from Balk et al. (*326*)]. The reactor conditions: TMG pressure, 83 mTorr; AsH$_3$ pressure, 3 Torr; total pressure, 75 Torr; laser energy fluence, 173 mJ/cm^2.

combining stable precursors with surface photons offers the real potential for obtaining useful growth rates of III–V films at reduced temperatures.

The influence of an ArF excimer laser beam ($\lambda = 193$ nm) on the GaAs growth rate as a function of temperature is displayed in Figure 61 for TMG as the Ga precursor. The transport-limited region of film growth is clearly noticeable for temperatures above $\sim 550°$C, but the optically induced increase in film growth rate rises rapidly at lower temperatures. Even larger enhancements are measured as the TMG partial pressure is increased (see Figure 62). The influence of surface reactions on this effect was borne out by comparing the spatial extent of the deposited film with the cross-sectional area of the laser beam.

In a series of experiments, Nishizawa et al. (*327–330*) explored the impact of excimer laser radiation at the substrate on the growth of GaAs by chloride transport vapor phase epitaxy (VPE) (*327–329*) and MOCVD (*330*). Not only were film growth rates enhanced, but the Hall mobilities of films grown in the presence of UV photons were improved over those deposited without substrate illumination. At a substrate temperature of 600°C, for example, the growth of epitaxial GaAs from AsCl$_3$ and GaCl$_3$ by VPE was roughly quadrupled (from ~ 1.2 to 4.5 μm·h^{-1}) by irradiating the GaAs(100) substrate with

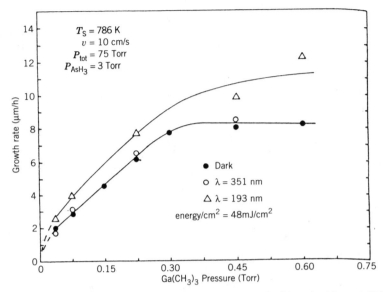

Figure 62. Increase in the GaAs growth rate that is measured with and without ArF laser illumination of the substrate [reprinted from Balk et al. (326), by permission]. The Ga-alkyl precursor is TMG, and the ArF laser energy fluence at the substrate is 48 mJ·cm^{-2}.

2.7 W of average laser power at 248 nm (KrF laser). Similar improvements were observed over the 480–700°C temperature range. Also, epitaxial films were grown at temperatures as low as 350°C. Although surface morphologies were noticeably improved when UV was present during growth, a dependence of the effect on wavelength (i.e., threshold wavelength) was not detectable (330). The influence of UV photons at the substrate on the electrical properties of the film (in addition to surface morphology) was investigated by Kukimoto et al. (331). In the ~600–800°C temperature range, it was found that irradiating the substrate with 6.4-eV (193-nm) photons had little influence on the film growth rate but that carrier concentrations increased in n-type regions and decreased in p-type regions. Furthermore, the concentration of Al in AlGaAs rose significantly in the presence of surface illumination. The latter effect is strongest at the lowest temperatures studied, with increases in the Al mole fraction (x) as large as ~0.03 observed at 600°C.

Chu et al. (332) grew epitaxial GaAs fims on n-type GaAs(100) substrates at temperatures in the 425–500°C range by irradiating the substrate during film growth with an ArF laser fluence of 19–38 mJ·cm^{-2}. In the absence of the UV irradiation, no deposition was observed at temperatures up to 600°C under conditions that were otherwise identical. Deposition rates ranging from

$0.02\ \mu\text{m}\cdot\text{min}^{-1}$ at 425°C to $0.1\ \mu\text{m}\cdot\text{min}^{-1}$ at 500°C were measured, and the electrical and chemical characteristics of the resulting films were determined to be similar to those for films grown by conventional MOCVD. Hole mobilities, for example, for films grown in the 425–500°C interval varied from 150 to 200 $\text{cm}^2/\text{V}\cdot\text{s}$, with carrier concentrations ranging from 5×10^{16} to $\sim 3 \times 10^{17}\ \text{cm}^{-3}$. The carbon number density was measured by SIMS to be 4×10^{17} to $2 \times 10^{18}\ \text{cm}^{-3}$ for films deposited at 500°C. York and co-workers (333, 334) carried out similar experiments but with lower laser fluences ($< 13\ \text{mJ}\cdot\text{cm}^{-2}$) and at wavelengths of 248 and 351 nm rather than 193 nm. They observed growth rate enhancements of 5–15% at 450°C when KrF photons were present, but no measurable effect was observed at 351 nm, which supports the photochemical nature of the enhancement. The effect was attributed to photolysis of adsorbed $Ga(CH_3)_3$ molecules: KrF was chosen rather than ArF to intentionally avoid photodissociation of AsH_3.

Atomic laser epitaxy (ALE) of GaAs was demonstrated by Kawakyu et al. (335) by illuminating the surface with 5 eV photons from a KrF laser. Instead of the continuous gas flow that is characteristic of most vapor phase film growth techniques, ALE entails pulsing the gas constituents sequentially. Kawakyu and co-workers, for example, grew GaAs in cycles where one cycle consisted of four gas pulses in the sequence: AsH_3, H_2, $Ga(CH_3)_3$, and H_2. Without the laser, self-limiting growth (i.e., one monolayer of GaAs per cycle) was observed in a ~ 30°C wide region centered at 495–500°C. As shown in Figure 63, however, the presence of surface photons accelerates the decomposition of TMG and its related fragments, allowing self-limited growth to be realized over a much wider temperature interval.

Virtually all of the photo-CVD growth of GaAs to date has involved direct illumination of the substrate during growth, but Ku et al. (336) recently grew GaAs on Si at 300°C with an ArF laser beam passing over the substrate. The composition of the films as determined by electron spectroscopy for chemical analysis (ESCA; see Figure 64) reveal little carbon and, although the results are preliminary, the films appear to be epitaxial with the (100)-oriented Si substrate.

Several groups have also successfully revisited the growth of GaAs films with UV lamps. In a clever series of experiments, Kachi et al. (337) combined a CW CO_2 laser with a high-pressure Hg lamp to grow epitaxial GaAs films at temperatures down to 500°C. The role played by the CO_2 laser was to vibrationally excite AsH_3 at the surface, while the Hg lamp served to improve layer quality and surface morphology. Confirmation of the CO_2 laser's involvement in the growth process was demonstrated by tuning the laser to the $10.531\ \mu\text{m}$ absorption line of AsH_3, whereupon growth rate enhancements > 2 were observed at 500°C. Tuning the CO_2 laser wavelength to a region

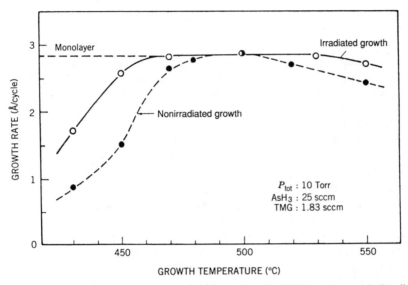

Figure 63. Growth rate of GaAs on Cr:GaAs(100) substrates by ALE for each gas cycle. Irradiating the surface with KrF photons gives a broad temperature range over which one monolayer/cycle (self-limiting) growth can be obtained [after Kawakyu et al. (*335*), by permission].

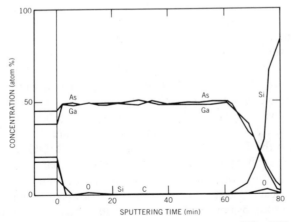

Figure 64. Depth profile of the composition of a GaAs film grown by photo-CVD as determined by ESCA. The film was grown in parallel geometry at $\sim 300°C$ [after Ku et al. (*336*)].

where AsH_3 does *not* absorb ($\lambda = 9.260\,\mu m$) or directing 10.531-μm radiation *parallel* to the substrate yielded no effect.

Norton and Ajmera (*338, 339*) deposited polycrystalline GaAs on quartz by photodissociating $Ga(C_2H_5)_3$ in the presence of AsH_3 with an Hg–Xe arc lamp (*338*) or by Hg photosensitization with a 500-W low-pressure Hg resonance grid lamp (10 sccm of H_2 through Hg bubbler). All of the films were deposited at 240°C, a temperature at which growth will not normally occur (i.e., without the lamp radiation).

A number of groups have demonstrated the growth of GaAs under conditions in which thermal effects are a significant, if not dominant, factor. Both visible Ar^+ (*340–346*) and pulsed excimer lasers (*80, 347–349*) have been used to produce spatially localized GaAs film growth by providing large intensities [50 to $> 1000\,W \cdot cm^{-2}$ continuously (*341–345*); $> 5\,MW \cdot cm^{-2}$ with pulsed sources] at the substrate. Despite the fact that this approach is relatively insensitive to the laser wavelength, Aoyagi et al. (*331–343*) and Bedair and co-workers (*344, 345*) have convincingly demonstrated that surface photochemical processes play a role in film deposition. Photo-assisted catalysis has been suggested (*343*) as a possible mechanism responsible for the observed acceleration in film growth rate.

Several recent studies have also shown the potential for photo-doping GaAs or *minimizing* the incorporation of impurities into GaAs with UV radiation. Iga and co-workers (*350*) found that irradiating the substrate with visible photons during the growth of GaAs by metal-organic molecular beam epitaxy (MOMBE) lowers the carbon impurity concentration by as much as 2 orders of magnitude. The effect is most pronounced at the lowest temperatures studied and is apparently the result of the 2.4-eV photons photolytically rupturing Ga—$CH_n (n \leqslant 3)$ bonds, culminating in the desorption of CH_n-containing species. Figure 65 illustrates the reduction in carbon concentration that was measured for films grown in the 400–500°C temperature range.

Intentionally *incorporating* carbon into GaAs has been accomplished by photodissociating CH_4 with a KrF laser (*351*). Carbon concentrations of $10^{21}\,cm^{-3}$ (in excess of the solid solubility limit), sheet resistances as low as $165\,\Omega/\square$, and surface carrier densities as high as $8 \times 10^{14}\,cm^{-2}$, were obtained. Enhanced Si doping of GaAs has also been demonstrated by introducing Si_2H_6 into the TEG/AsH_3 gas flow stream of an MOCVD reactor and illuminating the substrate with a high-pressure Hg–Xe lamp (*352*).

In GaAs, *p–n* junctions have been fabricated by simply shuttering the excimer laser beam. Specifically, Ban et al. (*353*) produced *p–n* junctions and doping superlattices by flowing the dopant precursors tetramethylsilane (TMSi) and dimethylzinc (DMZ) at a constant rate throughout the growth of GaAs films by MOCVD. Spatially modulating the Si concentration was accomplished by periodically blanking (turning off) the excimer beam. The 193-nm

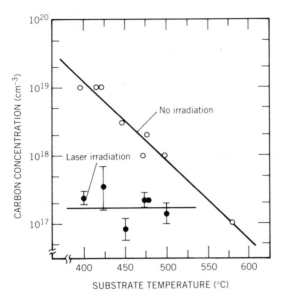

Figure 65. Carbon concentration in GaAs films deposited with visible laser ($hv = 2.4\,eV$) irradiation of the substrate (●) or in the absence of surface photons (○) [after Iga et al. (*350*), by permission].

photons from the laser interact primarily with TMSi, whereas at the growth temperature of 700°C, DMZ pyrolyzes rapidly and is not significantly impacted by the UV photons. Consequently, *n*-type layers are formed when the laser illuminates the substrate and *p*-type layers are grown when it is not. Junctions grown in this manner yield diodes having a breakdown voltage of $\sim 26\,V$ and a built-in voltage (V_0) of 0.8 V for a forward current of $1\,\mu A$. Doping superlattices were also fabricated, and Figure 66 gives the depth profile (recorded by SIMS) for a superlattice consisting of 650-Å periods. The significance of these and similar results obtained for Ge–Si superlattices (*48*) is that compositional and doping-modulated structures are now accessible by varying the intensity (or wavelength) of an optical source rather than by modulating the flow of gases with mass flow controllers. The former can be controlled electro-optically and has sub-microsecond turn "off" and "on" transients, whereas the latter are mechanical devices.

5.3.2. Al-Containing Compounds

Recently, Tokumitsu et al. (*80*) combined an ArF laser and a MOMBE reactor to grow epitaxial AlAs and polycrystalline GaAlAs films at 350°C. Crystalline

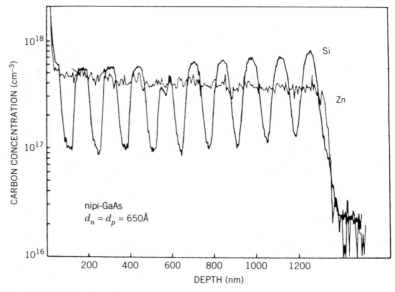

Figure 66. SIMS depth profile of a GaAs doping superlattice grown by photo-CVD at 700°C with an ArF laser [after Ban et al. (*353*)].

AlAs films up to 1.25 μm in thickness were grown by irradiating the substrate, whereas in the absence of UV photons films only ~ 500 Å thick were obtained. Similarly, 0.2-μm-thick polycrystalline GaAlAs films were deposited with surface illumination. Without the UV laser beam, film thicknesses were limited to 200–300 Å. For these studies, the gallium precursor was triethylgallium [$Ga(C_2H_5)_3$] and heating of elemental arsenic provided a beam of As_4 at the substrate. AlAs, GaAs, and [Al, Ga]As were deposited by Zinck and co-workers (*354*) from Lewis acid–base adducts such as $(CH_3)_3Al:As(CH_3)_3$. The vapor pressure of this adduct at 25°C is ~ 0.5 Torr (*354*), and both the Ga–As and Al–As adducts investigated in Ref. (*354*) absorb strongly at 193 nm but only weakly at 248 nm. Best film morphology was obtained when the excimer laser beam was directed parallel to the Ge substrate, but the maximum growth rates were observed to be 0.03 Å/pulse. For the GaAs films, the As/Ga ratio was found to increase for KrF laser fluences above 140 mJ/cm², and for AlAs and [Al, Ga]As films as well the film stoichiometry was observed to be a function of the laser fluence. The Ga/Al ratio in [Al, Ga]As films, for example, was determined to be 20 for $\lambda = 193$ nm, a laser fluence of 320 mJ/cm², and a Ga/Al ratio in the gas phase of 17. At 248 nm, however, a Ga/Al ratio (gas phase) of 14, and 100 mJ/cm² of fluence, the Ga/Al ratio in the deposit fell to 0.2.

5.3.3.   GaP, GaN

Epitaxial films of the wider bandgap binary GaP have been deposited by Sudarsan et al. (355–357) onto GaP(100) at 500°C by photodissociating a mixture of trimethylgallium and tertiary butylphosphine (TBP) with an ArF laser. In the absence of the excimer laser radiation, a thermal film growth rate of $60 \text{Å} \cdot \text{min}^{-1}$ is observed and the material is polycrystalline. With 110 mJ/cm^2 of 193-nm laser fluence at the substrate, however, epitaxial films are grown at a deposition rate six times larger (i.e., $\sim 360 \text{Å} \cdot \text{min}^{-1}$) than that of the thermal background. Although pulsed heating of the substrate undoubtedly plays a role in film growth for this fluence, identical experiments carried out at 248 nm yielded only polycrystalline films. Since both TMG and TBP absorb less strongly at 248 nm than at 193 nm, it is apparent that photodissociation of the precursors also plays a role in determining the structure of the film. Epitaxial films were obtained only in a narrow range of laser fluences. Below 100 mJ/cm^2 the films were polycrystalline, but above 120 mJ/cm^2 surface ablation dominated.

GaN films have been grown on GaAs(100) or basal plane (0001) sapphire substrates at temperatures as low as 625°C by photodissociating mixtures of TMG and NH_3 at 193 nm in parallel geometry (358). Growth rates up to $2 \mu\text{m} \cdot \text{h}^{-1}$ have been obtained, and X-ray analysis shows the films to be polycrystalline but preferentially oriented (0001) when grown at 625°C on basal plane sapphire substrates.

5.3.4.   InP and InSb

Indium phosphide (along with HgTe) was one of the first semiconductor compounds to be deposited by photochemical processes in the gas phase. Donnelly and co-workers (104, 359, 360) deposited amorphous, polycrystalline, and epitaxial InP films on quartz, GaAs, or InP substrates by photodissociating mixtures of $(CH_3)_3InP(CH_3)_3$ and $P(CH_3)_3$ with an ArF laser beam at normal incidence to the surface. Film thicknesses typically were $< 1000 \text{ Å}$, and deposition rates were measured to be $\sim 0.2 \text{ Å/pulse}$. Epitaxial films were obtained only for laser fluences exceeding $\sim 100 \text{ mJ/cm}^2$, which indicates the contribution of thermal effects to the deposition process. This was confirmed by the fact that film growth also occurred for an excimer laser wavelength of 351 nm (XeF), in a spectral region where both precursors are transparent. However, at 351 nm, the film growth rates were roughly a factor of 4 smaller than those at 248 and 193 nm, which suggests that although film growth is controlled by pyrolysis, photochemical mechanisms also contribute. The highest quality epitaxial films were grown at 320°C.

Zuhoski, Killeen and Biefeld (35) have recently reported the photodeposition of polycrystalline InSb films on GaAs(100) at room temperature. The precursors $In(CH_3)_3$ and $Sb(CH_3)_3$ both absorb strongly at 193 nm [see Table 1], but the absorption cross section falls by an order of magnitude in going to 248 nm. Nevertheless, σ at 248 nm for $In(CH_3)_3$ and $Sb(CH_3)_3$ is 1.2×10^{-18} and $1.7 \times 10^{-18}\,cm^2$, respectively (35), and it was found that the highest quality InSb films were deposited with KrF radiation (50 mJ/cm^2) and for an $Sb(CH_3)_3:In(CH_3)_3$ partial pressure ratio of 17:1.

5.4. SUMMARY

Photo-assisted reactions have proven to be capable of controlling the growth of epitaxial GaAs with surface photons and/or gas phase photochemistry. The effect of external optical radiation on growth rates, impurity incorporation, and carrier mobilities, for example, is most noticeable at low processing temperatures (< 600°C), where film growth is limited by surface reaction rates. Irradiating the surface with UV photons accelerates the surface decomposition of alkyls, in particular, and the desorption of contaminant-containing radicals. A dramatic demonstration of the utility of UV-stimulated reactions in film growth is the laser-assisted atomic layer epitaxy of GaAs. Although much of the effort on photo-assisted growth of III–V compound films has centered on GaAs, the extension of these techniques to GaP, GaN, and other binary and ternary compound films will become more prominent in the next few years.

The most challenging issue facing photo-epitaxy is tailoring the chemical and optical properties of precursors in such a way that, at the excitation wavelength, the decomposition and reaction pathway is selective. The unintentional incorporation of impurities into photo-CVD films can be avoided only by a detailed understanding of the processes by which precursor decomposition occurs. Carbon incorporation has been a problem in the growth of III–V and II–VI compound films that rely on alkyls as precursors, but several recent results in the parallel geometry growth of GaAs and the effective use of surface photons suggest that this difficulty can be minimized through the judicious choice of irradiation geometry, excitation wavelength, precursors, and gas partial pressures and flow rates in the reactor.

CHAPTER

6

DIELECTRICS

Of the materials that have been deposited by photo-CVD techniques, the dielectrics are among those that have progressed most rapidly toward commercial applications. Photo-CVD deposited SiO_2, for example, has excellent interface state density characteristics and low defect densities and has been successfully incorporated into InGaAs and InP MOS (metal-oxide semiconductor) devices (361, 362) and serves as a dielectric in the fabrication of 4-megabit memory devices (6, 10). Also, commercial reactors are now available for the broad-area deposition of both the oxides and nitrides of Si by Hg photosensitization and ArF laser photodissociation. In addition to SiO_2 and Si_3N_4, a variety of other dielectrics such as P_3N_5 and the oxides of aluminum, tin, and zinc have deposited by photochemical methods, and the growth conditions and film properties for each material will be reviewed here. Table 17 lists the properties of most of the dielectrics that have been photo-deposited and summarizes the growth conditions.

6.1. SILICON DIOXIDE

Silicon dioxide (SiO_2) is the most widely applied and thoroughly characterized of the photo-CVD dielectrics owing to its fundamental importance in VLSI technology (as a passivation layer material, in particular). It is also finding applications as a barrier against corrosion for metals in harsh chemical environments. The attractiveness of photo-deposited SiO_2 stems from its characteristic low processing temperatures ($< 300°C$), interface state densities, and the absence of heavy ion damage. Combined with its step coverage and capacity for broad area deposition, these features make this dielectric of increasing interest for incorporation into commercial semiconductor processing. Photodepositing SiO_2 entails irradiating a mixture of a silane (SiH_4, Si_2H_6, or Si_3H_8) and an oxygen precursor (N_2O, NO_2, or O_2) with a UV or VUV optical source. Excellent results have been obtained with Hg or D_2 lamps, and detailed studies of the structural and electrical characteristics of photo-CVD SiO_2 films have shown them to be comparable in quality or superior to plasma-CVD films.

143

Table 17. Summary of Conditions under Which Dielectric Films Have Been Deposited by Photo-CVD

Compound	Precursors	Optical Source	λ(nm)	Experimental Conditions	Refs.	Comments
Al_2O_3	$Al(CH_3)_3$, N_2O	ArF, KrF lasers	193, 248	Si, III–V wafers or glass substrates at 350°C; N_2O mass flow rate: 50–1000 sccm; total reactor pressure: 1 Torr; ∥ geometry, laser beam 1 mm above substrate	194	Deposition rates of 2000 Å·min^{-1} were obtained at 248 nm and an average laser power of 10 W; at 193 nm and 8.2 W of average power, uniform film thickness required reducing TMA pressure to 30 mTorr. Best films were obtained with 200–800 sccm N_2O and 80–120 mTorr TMA.
Al_2O_3	$Al(CH_3)_3$, N_2O	ArF, KrF lasers	193, 248	Si substrates; substrate temperature: 300–500°C; TMA: N_2O pressure ratio: 1:50 for ArF and 1:17 for KrF; 0.5 Torr total pressure; average laser power: 3 W, 2 Hz (193 nm), 5 W (248 nm); parallel geometry	390	The influence of surface irradiation on Al_2O_3 films deposited by photo-CVD was investigated for surface fluences between <1 and 30 mJ·cm^{-2}; even low fluences increase film refractive index.
Al_2O_3	Diisopropoxy aluminum diketonates	Hg–Xe arc lamp (1 kW), Ar^+ (SH)^a	UV continuum, 257	36% of Hg–Xe power lies in 230–400 nm region; GaAs substrate; temperature: ⩽ 325°C	392	Film deposition rates of 0.25 $\mu m \cdot h^{-1}$ were measured with $Al(OPr^i)_2 Acac(OMe)$ as the precursor and the Hg–Xe lamp; with the 257 nm laser beam, deposition rates were ∼ 100 Å·h^{-1}. Below 200°C, Al_2O_3 films were amorphous.
Cr_2O_3, CrO_2	CrO_2Cl_2	Ar^+ laser	488–515	Si, SiO_2, GaAs, and glass substrates at room temperature; chromylchloride pressure ∼ 0.1 Torr; laser power: 1 mW to several W; 5 μm beam diameter at substrate; ⊥ geometry	394	Lines composed of Cr_2O_3 and CrO_2 were deposited on Si(100) at a scan rate of 20 $\mu m \cdot s^{-1}$ with 0.8 W of laser power (488 nm); CrO_2 single crystals up to ∼ 1 mm in length were grown on fused silica under the same conditions; process is initiated by the single photon photolysis of adsorbed chromylchloride, but thermal processes become dominant as the growing film absorbs more than ∼ 10–20 mW of laser power.

144

Material	Reactants	Lamp	Wavelength	Conditions	Ref.	Results
P_3N_5	PCl_3, NH_3, H_2	Hg lamp	185 nm	Substrate temperature: 250°C; mass flow rates: PCl_3, 30 sccm; NH_3, 60–120 sccm; total pressure: 5 Torr	398	At 250°C, deposition rate was 12.5 Å·min⁻¹; diodes fabricated from 50 Å thick P_3N_5 films on n-InP showed reverse saturation currents as low as 0.1 nA, breakdown voltages of 30 V, and a Schottky barrier height of 0.81 eV.
SiO_2	SiH_4, N_2O (+Hg vapor)	Hg lamp	254 nm	50–200°C deposition temperature; total reactor pressure: 0.3–1.0 Torr (see Tables 4 and 5)	177	Deposition rates up to 150 Å·min⁻¹ have been observed (typical value: 30 Å·min⁻¹); thickness uniformity of ±3% has been achieved (see Tables 5 and 18).
SiO_2	Si_2H_6, O_2	Hg lamp (low pressure, 110 W)	185, 254	1% Si_2H_6 in He, 200 sccm mass flow rate; O_2, 600 sccm; N_2 carrier, 1500 sccm; substrate temperature: 150–350°C	367	Direct photochemical deposition of SiO_2 was observed (i.e., in absence of Hg); deposition rates of > 500 Å·min⁻¹ were observed for photo-CVD at 300°C but maximum rates of ~350 Å·min⁻¹ for the thermal process at the same temperature.
SiO_2	SiH_4, O_2	Hg–Xe arc lamp (high pressure, 200 W)		4% SiH_4 in Ar: 50–70 sccm; O_2: 10–14 sccm; volumetric gas ratio (O_2:SiH_4), 5:1; total reactor pressure: 2 Torr; Si substrates, 150–350°C; typical deposition time: 11 min	147	Deposition rates measured at 150, 250, and 350°C were 230, 680, and 1140 Å·min⁻¹ (averaged over 3 trials); refractive index of films: 1.44–1.46.
SiO_2	SiH_4, O_2	Hg lamp (low pressure)	185, 254	N_2 carrier gas; N_2:SiH_4:O_2 ratio of 50:1:10 at a total reactor pressure of 5 Torr; lamp intensity: 4 mW·cm⁻² at 185 nm and 35 mW·cm⁻² at 254 nm; InP substrate, 51 mm dia., temperatures between 150 and 300°C.	368	Deposition rate at 300°C was ~1000 Å·min⁻¹ and the process activation energy is 0.56 eV. Deposition was shown to arise only from the 185 nm line of the lamp.
SiO_2	SiH_4, N_2O	Hg lamp	185, 254	SiH_4:N_2O gas ratio: 1:50–200; total pressure: 30 Torr; substrate temperature: 350–400°C	6	Deposition rates of 200 Å·min⁻¹ were observed, and the refractive index of the films was 1.46.
SiO_2	$Si(OC_2H_5)_4$	Hg lamp (low pressure)	185, 254	Mixture of tetraethoxysilane, acetone, and collodion spin-coated onto n-type Si substrate; time of exposure to optical source: 1 h; lamp intensity: 3 mW·cm⁻² (254 nm).	370	Films up to 2000 Å thick were obtained with refractive indices between 1.43 and 1.45.

145

Table 17 (*Continued*)

Compound	Precursors	Optical Source	λ(nm)	Experimental Conditions	Refs.	Comments
SiO_2	Si_3H_8, O_2	D_2 lamp (150 W)	115–400 (peak ~160 nm)	Si substrate, 25–390°C; mass flow rates: 0.11 sccm Si_3H_8, 0.65 sccm O_2; total pressure: 0.2 Torr	372	Maximum deposition rate, obtained at 25°C, was 150 Å·min⁻¹ and growth slowed with increasing temperature.
SiO_2	Si_2H_6, Si_2F_6, O_2	D_2 lamp (150 W)	115–400	Si substrate, 200–300°C substrate temperature; mass flow rates: Ar (carrier), 150 sccm; O_2, 15 sccm; Si_2H_6, 3.6 sccm; Si_2F_6, 0–3.5 sccm	375	Improved SiO_2 film deposition rates and reductions in Si—OH and Si—H defects were observed by adding Si_2F_6 to Si_2H_6/O_2 gas stream.
SiO_2	SiH_4, N_2O	Windowless N_2 lamp		Si(100) substrates, 275°C substrate temperature; mass flow rates: 70 sccm (N_2O), 1 sccm (SiH_4); total reactor pressure: ~0.8 Torr; lamp intensity: 0.1 mW·cm⁻² at substrate.	189	Deposition rates of ~180 Å·min⁻¹ were obtained, and thin film transistors were fabricated with photo-CVD SiO_2 as the gate dielectric.
SiO_2	SiH_4, NO_2	Ar resonance lamp	106.6	Si substrates, 25–250°C substrate temperature; mass flow rates: 40 sccm (NO_2), 3 sccm (SiH_4); NO_2 partial pressure: 0.25 Torr	378	100 Å·min⁻¹ deposition rates were obtained on Si; higher rates were obtained for HgCdTe. Films deposited at 250°C had a refractive index of 1.45–1.47.
SiO_2	SiH_4, N_2O	ArF laser	193	Average laser intensity: 40 MW·cm⁻²; peak intensity: 40 MW·cm⁻² at 100 Hz repetition frequency; ∥ geometry; reaction volume cross-sectional area: 1.5 mm × 12 mm; mass flow rates: 70 sccm (5% SiH_4 in N_2), 800 sccm (N_2O); total pressure: 8 Torr; Si(100) substrates	379,380	Films deposited below 200°C were milky in appearance; above 200°C, films were transparent and uniform; deposition rates up to 3000 Å·min⁻¹ were obtained, and between 250 and 350°C the refractive index ranged from 1.476 to 1.457.
SiO_xN_y	Si_2H_6, NH_3, NO_2	D_2 lamp		Si(100) substrate, 330°C substrate temperature; VUV intensity at substrate ($\lambda \leq 180$ nm): 12 mW·cm⁻²; mass flow rates: 2.7 sccm (Si_2H_6), 130 sccm (NH_3), 0.03–0.3 sccm (NO_2) and 500 sccm (N_2 carrier); total pressure: 11 Torr	382	Oxynitride films were deposited at rates up to 120 Å·min⁻¹ at 330°C; film composition is sensitive to NO_2 mass flow rate.

Material	Precursors	Source	Ref.	Conditions	Ref.	Comments
$Si_xO_yN_z$	SiH_4, NH_3 (+Hg vapor)	Hg lamp	254	100–200°C deposition temperatures; total reactor pressure: 0.3–1.0 Torr [see Tables 4 and 5 for parameters from Schuegraf (177)]	177,384 385	Deposition rates up to 65 Å·min^{-1} (typical value: 30 Å·min^{-1}) have been measured; refractive indices ranging from 1.8 to 2.4 are obtained by altering the gas mixture (see Tables 5 and 18).
SiN_x	Si_2H_6, NH_3	Hg lamp (low pressure)	185,254	Si_2H_6:NH_3 volume ratio: 1:50; substrate temperature: 300–350°C; total reactor pressure: 3 Torr	6	A deposition rate of 40 Å·min^{-1} was observed, and the refractive index of the films was 2.0.
Si_xN_y	SiH_4, NH_3	ArF laser	193	Si substrate, 200–425°C substrate temperature; ∥ geometry; mass flow rates: 10 sccm NH_3, 10 sccm SiH_4, 50 sccm N_2; total pressure: 2 Torr	387	Deposition rates up to 700 Å·min^{-1} were observed, and films are comparable to plasma-deposited nitrides.
Si_3N_4	Si_2H_6, NH_3	ArF laser	193	Si substrate, 25–550°C substrate temperature; ⊥ geometry; laser fluence: ~15 mJ·cm^{-2}; mass flow rates: 6 sccm (Si_2H_6 in He), 250 sccm (NH_3); total pressure: 0.6 Torr	388	Below 400°C, growth is primarily photochemical at a rate of 50 Å·min^{-1}; at 550°C, the thermal deposition rate is ~200 Å·min^{-1}; film composition varies with substrate temperature.
SnO_2	$SnCl_4$, N_2O	ArF laser	193	Fused silica substrates at room temperature; 10–20 mJ·cm^{-2} per pulse laser energy fluence; exposure: typically 10^4 pulses, 5 Hz laser repetition frequency; 0.75 Torr $SnCl_4$, 50 Torr N_2O; ⊥ geometry	393	1000 Å thick films were deposited that are conducting for laser fluences of 10–20 mJ·cm^{-2}; at lower fluences, only $SnOCl_2$ is formed; beyond 70–80 mJ·cm^{-2}, ablation dominates.
TaO_x	$TaCl_5$, O_2	Hg lamp (low pressure)	185,254	n-Type Si(100) substrates; 100–500°C substrate temperature; $TaCl_5$ source at 120°C; 200 sccm O_2, 10 sccm N_2; total pressure: 1–7 Torr	125	Deposition rates exceeding 30 Å·min^{-1} were observed (1 Torr total pressure), but in the absence of UV radiation the deposition rate was only 2 Å·min^{-1} at 400°C; deposition rate also increases with decreasing pressure.

Table 17 (*Continued*)

Compound	Precursors	Optical Source	λ(nm)	Experimental Conditions	Refs.	Comments
TaO_x	$Ta(OCH_3)_5$, O_2 or N_2	Hg lamp	185, 254	n-Type Si(100) substrates; 150–400°C substrate temperature; $Ta(OCH_3)_5$ source at 100–130°C; mass flow rates: 50–200 sccm O_2, 10 sccm N_2; total pressure: 1.5 Torr; lamp intensity: 30–40 mW·cm^{-2} at substrate (254 nm); < 5 mW·cm^{-2} (185 nm)	126	Amorphous TaO_x films were deposited at a rate of 120 Å·min^{-1} at 350°C; X-ray measurements show the films to be a mixture of Ta_2O_5 and TaO_2.
TiN	$TiCl_4$, NH_3 (or N_2)	D_2 lamp (150 W)		Stainless steel substrates; mass flow rates: 0.2 sccm $TiCl_4$, 3.1 sccm N_2, 5.1 sccm H_2 (carrier); reaction time: 1 h; substrate temperature: 650–1000°C	397	At 900°C, the TiN film deposition rate for 0.2 sccm $TiCl_4$ is 3 times larger than that obtained without irradiation; the enhancement at 800°C is 35%; irradiation permits the film growth temperature to be lowered by 50–100°C.
VO_x	$VOCl_3$	Hg lamp (high pressure)		Vycor glass substrates at 0°C·	395	Vanadium oxide catalysts were formed by the reaction of $VOCl_3$ with OH groups on the substrate surface in the presence of UV radiation.

148

6.1.1. Hg Photosensitization

The deposition of SiO_2 by Hg photosensitization was demonstrated in the early 1980s (177,363). Irradiating mixtures of SiH_4, N_2O, and Hg vapor yielded deposition rates up to $150 \, \text{Å·min}^{-1}$ at a total pressure of 0.3 to 1.0 Torr and substrate temperatures as low as 50°C. Tables 4, 5, and 18 summarize the reactor conditions and electrical and structural characteristics of the deposited dielectric films (177,364). More recent studies with a commercially available reactor examined silicon oxide deposition at 100 and 200°C (365). Deposition rates in excess of $150 \, \text{Å·min}^{-1}$ were observed, but pinhole densities ranging from $\sim 200/\text{cm}^2$ to $600/\text{cm}^2$ and refractive indices between ~ 1.5 and 2.0 were measured for 0.3-μm-thick films grown at 200°C. Depending on the growth conditions, intermediate oxide phases such as Si_2O_3, SiO, and Si_2O were also present. Infrared spectroscopy of films deposited by Hg photosensitization shows that, by controlling the SiH_4 partial pressure, films can be grown in the 100–160°C range having properties similar to those for an SiO_2 film deposited thermally at 1100°C.

With InP MISFET (metal-insulator-semiconductor field effect transistor) applications in mind, Su et al. (362) characterized in detail the properties of photo-CVD SiO_2 films deposited onto InP. As is characteristic of photo-CVD processes, the presence of UV photons in the reactor significantly reduces

Table 18. Characteristics of Silicon Oxide and Silicon Nitride Films Deposited by Photo-CVD in a Commercial Reactor

Characteristic	Film	
	Silicon Dioxide	Silicon Nitride
Density	$2.10 \, \text{g/cm}^3$	1.8 to $2.4 \, \text{g/cm}^3$
Index of refraction	1.45 to 1.49[a]	1.8 to 2.4[a]
Dielectric constant	3.9	5.5
Dielectric strength	$4.0\text{–}6.0 \times 10^6 \, \text{V·cm}^{-1}$	$4 \times 10^6 \, \text{V·cm}^{-1}$
Defect density	$< 10/\text{cm}^2$	$< 10/\text{cm}^2$
Film adhesion (tensile)	$> 5.2 \times 10^5 \, \text{Torr} \, (10^4 \, \text{lb/in.}^2)$	$3.1\text{–}5.2 \times 10^4 \, \text{Torr}$ $(0.6\text{–}1.0 \times 10^3 \, \text{lb/in.}^2)$
Etch rate	$9 \, \text{nm·s}^{-1}$ [b]	$6\text{–}10 \, \text{nm·s}^{-1}$ [c]
Film composition	SiO_2[a]	$Si_xO_yN_z$[a]

Source: After Schuegraf (177).

[a] Film composition characteristics can be varied by reactant gas ratio.
[b] Buffered oxide etch, film densified (15 min at 450°C in N_2).
[c] 1:10 HF, as deposited.

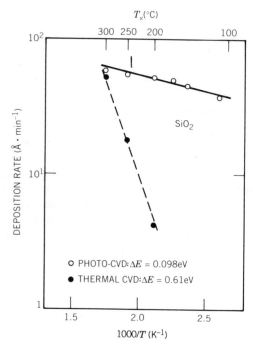

Figure 67. SiO$_2$ deposition rate as a function of substrate temperature for photo-CVD (○) and thermal CVD (●) films [after Su et al. (*362*)]. Note that the activation energy for the photochemically deposited films is less than 0.1 eV.

the dependence of the film growth rate on substrate temperature. Figure 67 shows the data for films deposited in the 100–300°C range (*362*). Conventional thermal deposition (CVD) is characterized by an activation energy, E_a, of 0.61 eV (denoted by the dashed line in Figure 67), whereas for photo-CVD E_a is less than 0.1 eV. By varying the ratio of the SiH$_4$ to N$_2$O mass flow rates, the composition of the deposited films can, as illustrated in Figure 68, be altered from SiO$_2$ to SiO$_x$ ($x < 2$). The refractive index for stoichiometric silicon dioxide is 1.46, while the higher indices of refraction in Figure 68 are expected for films having excess Si.

Because of the low processing temperatures, sharp overlayer/substrate interfaces result. The Auger depth profile for an SiO$_2$ film photodeposited onto InP (Figure 69) shows an SiO$_2$/InP interfacial region roughly 65 Å in width, and the Hg concentration in the SiO$_2$ film is below the detectability limit of 0.1% (*362*). Al/SiO$_2$/InP MOS devices were fabricated from the photo-CVD films (*362*), and the capacitance–voltage (*C–V*) characteristic (see

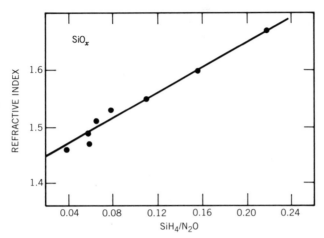

Figure 68. Variation of the refractive index of photo-CVD SiO_x films with the SiH_4/N_2O gas flow ratio. The refractive index of SiO_2 films is approximately 1.46 [after Su et al. (*362*)].

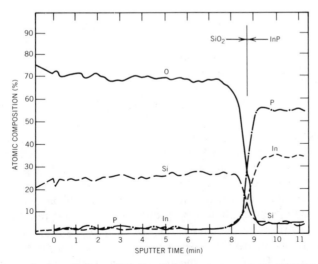

Figure 69. Auger depth profile for an SiO_2 film photodeposited onto InP at 250°C [after Su et al. (*362*)].

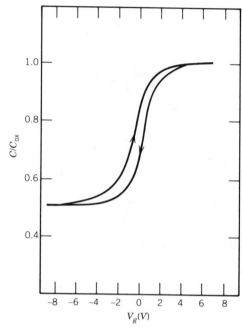

Figure 70. $C-V$ characteristic at 1 MHz for an Al/SiO$_2$/InP MOS device in which the SiO$_2$ layer was photodeposited at 250°C [after Su et al. (*362*), by permission].

Figure 70) displays hysteresis and yields an estimated oxide charge density of 4.5×10^{11} cm^{-2}. Also, the interface trap density shown in Figure 71 has a minimum value of 1.2×10^{11} cm$^{-2} \cdot$eV^{-1}, and the breakdown strength of the dielectric was measured to be 7.0×10^{6} V/cm.

6.1.2. Direct Photodecomposition

Despite the simplicity of the growth of SiO$_2$ (or Si$_3$N$_4$) by Hg photosensitization, contamination of the films and the toxicity of Hg itself are distinct drawbacks of the process that (as was the case in the photodeposition of Si) have prompted the photo-CVD community to pursue direct photodecomposition of the reactants with lamps or a laser. Mishima and co-workers (*367*) reported the direct photodeposition of SiO$_2$ from mixtures of SiH$_4$ and O$_2$ with a low-pressure mercury lamp. For each temperature studied in the 150–350°C range, photo-CVD deposition rates exceeded those obtained by pyrolysis. At 300°C, the photochemical and thermal deposition rates were > 500 and ~ 350 Å·min^{-1}, respectively. Also, since the activation energy for the photo-CVD process in the 150 to ~ 280°C range (0.53 eV) is roughly half

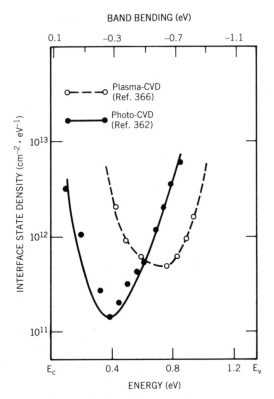

Figure 71. Comparison of the interface trap density for the Al/photo-SiO$_2$/InP MOS device of Figure 70 with that for an analogous structure fabricated by plasma-CVD [after Su et al. (*362*) Pande and Nair (*366*)].

that for thermal deposition ($E_a = 0.96$ eV) (*367*), the difference between the photo and thermal deposition rates becomes more pronounced at lower substrate temperatures. At 150°C, the rates differ by more than an order of magnitude. Also utilizing a low-pressure Hg lamp, Nissim and co-workers (*368*) deposited SiO$_2$ onto InP substrates at rates up to 1000 Å·min^{-1} at 300°C. It was demonstrated that deposition arose from only the 185 nm line of the lamp and the activation energy for the process was measured to be 0.56 eV, nearly identical to the value reported in ref. (*367*). For a substrate temperature of 200°C, films having a refractive index of 1.47 were grown at a rate of 125 Å·min^{-1}. When deposited on undoped InP substrates, the resistivity of the insulating film was 9×10^{14} Ω·cm and its breakdown voltage, 6×10^6 V/cm, is comparable to that for films deposited by Hg photosensitization. Petitjean et al. (*369*) also photodeposited SiO$_2$ onto InP and Si at temperatures between 50 and 300°C by photolyzing SiH_4/N_2O mixtures with an Hg lamp (185 nm). Deposition rates varied from 7 to 20 Å·min^{-1} and refractive indices between 1.535 and 1.56 were measured. Breakdown fields as high as 9 MV·cm^{-1} were obtained. Also, the $C–V$ characteristics of MIS diodes fabricated from the photo-CVD films (by evaporating Al dots onto the dielectric) were measured at 1 MHz. For SiO_2 on n-InP, the interface charge density determined from the flat band voltage was 1.1×10^{11} cm^{-2}. Although not a gas phase photochemical process, the technique reported by Niwano et al. (370) is a clever one in that it relies on tetraethoxysilane $[Si(OC_2H_5)_4]$ spun onto a substrate to yield SiO_2 directly.

Scoles et al. (148) carried out similar experiments with monosilane and O_2 with an Hg–Xe arc lamp acting as the optical source. For substrate temperatures between 150 and 350°C, deposition rates ranging from 230 to 1140 Å·min^{-1} were recorded and the electrical and structural properties of the films (such as the dielectric constant, breakdown voltage, and capacitance) were found to be comparable to those for thermally deposited films.

Several groups have also explored the D_2 lamp as an excitation source (371–373). Silicon dioxide films were deposited by Okuyama et al. (372) from Si_3H_8/O_2 mixtures at a rate of 150 Å·min^{-1} at room temperature (25°C). The deposition rate declined with increasing temperature, which was attributed to desorption of an adlayer of lamp-produced radical species. Subsequent experiments by Okuyama and co-workers (374) involved both a 150-W D_2 lamp *and* a 250-W mercury lamp to deposit SiO_2 from Si_3H_8/O_2 mixtures. The largest deposition rate—150 Å·min^{-1}—was again recorded at 25°C, but measurements of the optical and electrical characteristics of the dielectric films demonstrated that the best results require higher temperatures. The index of refraction of the films reached 1.452 for a substrate temperature of 160°C and fell monotonically for both lower and higher temperatures. MOS diodes were fabricated from 1000-Å-thick photo-CVD films deposited on n-Si(100) wafers having a resistivity of 2.9 Ω·cm. $C–V$ measurements made at 1 MHz yield a fixed oxide charge density of $\sim 2 \times 10^{11}$ cm^{-2} for films grown at 280°C. This result and the measured breakdown fields ($\leqslant 7$ MV·cm^{-1}) are comparable to the characteristics of films photodeposited with a 185-nm Hg lamp (described earlier). Using deep level transient spectroscopy (DLTS), Okuyama et al. (374) were also able to determine the interface state densities for various films. The lowest density, 3.6×10^{10} cm^{-2}·eV^{-1}, was measured at $E–E_c = 0.4$ eV for a film grown at a substrate temperature of 260°C.

Nonaka and co-workers (375) have reported significant increases (i.e., factors of 2–4) in the deposition rate of amorphous SiO_2 films by adding Si_2F_6 to the feedstock gas stream. At 200°C and an Si_2H_6 mass flow rate of 2.4 sccm, for example, the SiO_2 deposition rate rose linearly from

~ 140 Å·min^{-1} to almost 600 Å·min^{-1} by introducing ~ 3.5 sccm of Si$_2$F$_6$ into the gas stream. Similar improvements in deposition rate (factors of 2–3) were also observed at other disilane mass flow rates. Perhaps more importantly, the presence of Si$_2$F$_6$ (or SiF$_4$) dramatically reduced the number of Si—OH and Si—H defects in the dielectric film. An order of magnitude decrease in the interface state density (N_{SS}) in SiO$_2$ films photodeposited onto Si from Si$_2$H$_6$/O$_2$ mixtures with a D$_2$ lamp has also been realized by treating the substrate with F$_2$ (376). Figure 72 portrays the reduction in N_{SS} that occurs if F$_2$ (diluted in He) is flowed through the reactor immediately prior to the photodeposition of SiO$_2$. The minimum interface state density is obtained for a substrate temperature of 230°C. Now, if the substrate is also irradiated during the F$_2$ treatment phase with UV photons *from a second lamp*, further reduction in N_{SS} occurs (Figure 72). Infrared spectra show that this effect arises from a reduction in the number of Si—H bonds at the surface (377).

Efforts to deposit SiO$_2$ with shorter wavelength optical sources include the work of Baker, Milne, and Robertson (189) with a windowless N$_2$ lamp.

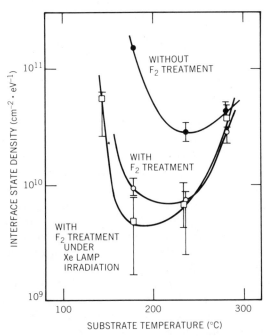

Figure 72. Interface state densities (N_{SS}) observed for SiO$_2$ films photodeposited onto Si without (●) and with (○) F$_2$ treatment of the substrate. If the F$_2$ is irradiated with the UV output of a Xe lamp during the substrate treatment (□), further reduction in N_{SS} is observed [after Inoue et al. (376)].

At 275°C, the deposition rate was $180 \text{Å} \cdot \text{min}^{-1}$ and the refractive index of the films was measured to be 1.46. Although the film deposition process is initiated by lamp radiation, atomic nitrogen produced in the lamp also appears to play a significant role. Thin film MOS field effect transistors fabricated from such photo-CVD, amorphous SiO_2 films yielded a gate leakage current of 1 pA. Silicon dioxide films have also been deposited with VUV resonance radiation at 106.6 nm from a microwave-excited Ar lamp. Marks and Robertson (378) photodissociated mixtures of NO_2 and SiH_4 at 100°C (and up to 250°C) with the lamp to obtain deposition rates of $100 \text{Å} \cdot \text{min}^{-1}$ on silicon and the resulting films exhibited refractive indices between 1.45 and 1.47. Remarkably, the lamp's optical flux at a distance of 4 cm was $3 \times 10^{14} \text{photons/cm}^2 \cdot \text{s}$.

While most SiO_2 deposition studies have relied on lamps as the optical source, Boyer et al. (379, 380) resorted to an ArF (193-nm) laser in parallel geometry to deposit SiO_2 films at rates up to $3000 \text{Å} \cdot \text{min}^{-1}$. The deposition rate was insensitive to substrate temperature in the 20–600°C temperature interval, but was linear in gas pressure and laser intensity. Transparent, scratch-resistant films were obtained at temperatures above 200°C, and the measured refractive indices ranged from 1.476 (250°C) down to 1.457 (350°C). Studies reported by Szorényi et al. (381) with a similar arrangement (parallel geometry) also utilized an ArF laser operating at 200 Hz and producing an average power of 35 W. In agreement with the results of refs. 370 and 371, the film growth rate rose linearly with the laser energy fluence $(\text{mJ} \cdot \text{cm}^{-2})$, but the index of refraction of the deposited films could be varied over the range of ~ 1.4–2.1 by varying the N_2O or SiH_4 partial pressures. Tsuji, Itoh, and Nishimura (429) obtained stoichiometric SiO_2 films for growth temperatures in the 150–400°C range by also photodissociating SiH_4/N_2O mixtures in parallel geometry with an ArF laser. Films 2000 Å thick were grown at rates up to $70 \text{Å} \cdot \text{min}^{-1}$ and depositing films of the proper stoichiometry required that the $N_2O:SiH_4$ ratio be in the range 400–1200.

Oxynitrides of Si have been deposited by adding ammonia to a disilane and nitrogen dioxide (NO_2) gas stream. Watanabe and Hanabusa (382) showed that the SiO_xN_y composition is quite sensitive to the NO_2 mass flow rate, which is the smallest of those for the three precursors. Deposition rates up to $120 \text{Å} \cdot \text{min}^{-1}$ were reported.

On the basis of extensive studies of the photo-CVD of SiO_2 on various substrates, it is fair to conclude that films having properties comparable to or superior to films deposited by conventional means (thermal or plasma CVD) can now be obtained. Specifically, breakdown voltages of > 7–9 $\text{MV} \cdot \text{cm}^{-1}$, fixed oxide charge densities in MOS devices of 1–$2 \times 10^{11} \text{cm}^{-2}$ and interface state densities $N_{SS} < 10^{11} \text{cm}^{-2} \text{eV}^{-1}$ are now routinely available. Not surprisingly, surface studies of the photo-CVD process (383) reveal that the species deposited on the substrate are similar to those found when thermal deposition occurs. Thorough characterization of photodeposited films with regard to growth rates, refractive index and electrical characteristics confirms the potential of photo-CVD SiO_2 as a dielectric in the low-temperature fabrication of MOS devices.

6.2. SILICON NITRIDE

Although photodeposited silicon nitride (Si_3N_4) has not been as thoroughly characterized as its oxide counterpart, the growth conditions for the two dielectrics are similar. Several groups (177,384,385) recognized that Hg photosensitization of SiH_4 in the presence of NH_3 would yield silicon nitride films. Utilizing a process developed by Hughes Aircraft, Schuegraf (177) has obtained deposition rates up to $65\ \text{Å·min}^{-1}$ in the 100–200°C temperature range, with Hg impurity concentrations in the resulting films of several parts per billion. Also, the film index of refraction is adjustable between 1.8 and 2.4 by varying the gas mixture ratio. Further detail regarding the deposition process parameters and film characteristics can be found in Tables 4, 5, and 18.

Inushima et al. (6) photodeposited SiN_x films by photodissociating NH_3 and Si_2H_6 (in the absence of mercury vapor) with the 185 nm line from an Hg resonance lamp. For substrate temperatures between 300 and 350°C, the deposition rate was $40\ \text{Å·min}^{-1}$, the hydrogen content was 20 atom %, and the refractive index for the films was 2.0. Also, the deposition rate was independent of the substrate temperature and the film thickness was uniform to $\pm 5\%$. By varying the distance from the lamp to the substrate, Inushima and co-workers (386) subsequently determined that, under their operating conditions, the migration length of the active species in the SiN_x film growth process is 10–20 mm.

Early demonstrations of the ability of excimer laser radiation to deposit silicon nitride were carried out by Boyer et al. (387). With the ArF laser beam parallel to the substrate, they obtained deposition rates up to $700\ \text{Å·min}^{-1}$ and the physical properties of the photodeposited films were found to be comparable to those for plasma-CVD films but of higher quality (particularly insofar as pinhole densities are concerned) than films deposited by Hg photosensitization. Weak illumination of the substrate with 254- or 193-nm photons ("surface photons") reduced the etch rate of the laser-deposited nitride films by more than a factor of 5. Similar results were obtained by Sugii, Ito and Ishikawa (388), who explored the deposition of Si_3N_4 by photodissociating mixtures of Si_2H_6 and NH_3 with an ArF laser in *perpendicular* geometry. Below 400°C, film growth, as portrayed in

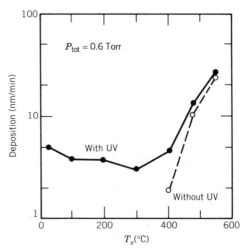

Figure 73. Dependence of the deposition rate of Si_3N_4 on temperature: (a) with ArF laser irradiation of the substrate (●), and (b) in the absence of UV irradiation (○). Note that the ordinate of the graph is logarithmic and the total reactor pressure is 0.6 Torr [after ref. (388)].

Figure 73, is primarily photochemical and the deposition rate is roughly constant at 30–$50\,\text{Å}\cdot\text{min}^{-1}$. Thermal effects become dominant at higher temperatures, and at 550°C the deposition rate has risen to $\sim 200\,\text{Å min}^{-1}$. AES reveals that obtaining stoichiometric silicon nitride is hindered by the presence of oxygen in most of the films reported to date. Specifically, the composition of the photodeposited films is determined to be $Si_3O_xN_y$, where the nitrogen content gradually declines for substrate temperatures above 200°C, with y falling from ~ 4 at 200°C to ~ 3.5 at 400°C. Infrared absorption spectra of films deposited at 350°C show weak features attributable to N—H bonds, but Si—H absorption bands were not detectable. It appears that hydrogen atoms, produced in the gas phase or on the surface by the dissociation of Si_2H_6 or NH_3 [i.e., $NH_3 + h\nu \rightarrow NH_3(A) \rightarrow NH_2(X) + H$, etc.], selectively attack and eliminate Si—H bonds on the surface of the growing film. In agreement with ref. (387), it was found that direct irradiation of the substrate during deposition improved the film's etch rate characteristics. Silicon nitride films deposited at 400°C have an etch rate of $15\,\text{Å}\cdot\text{min}^{-1}$, which is comparable to that for films grown by conventional (thermal) CVD at 800°C. Recently, Urisu et al. (190) have deposited hydrogenated silicon nitride films at 190°C with radiation from a synchrotron. Since VUV and extreme ultraviolet (XUV; $\lambda \leqslant 100\,\text{nm}$) photons are available with this device, precursors having stronger chemical bonds (i.e., more stable) are accessible and N_2, rather than ammonia, can be used as the nitrogen precursor. With a mixture composed of 0.02-Torr SiH_4 and 0.1-Torr N_2, deposition rates of 4 Å·min^{-1} were obtained for a ring current of 100 mA.

Few characterizations of the electrical properties of silicon nitride films have appeared in the literature as most efforts on this material have focused thus far on parameterizing the growth process. However, Iborra et al. (389) have examined silicon nitride films grown by Hg photosensitization on Si(100) at temperatures varied from 70 to 200°C. Deposition rates ranging from 15 to 40 Å·min^{-1} were measured and MIS devices were fabricated by depositing 1000 Å of silicon nitride on n-Si substrates having a doping concentration of $7 \times 10^{14} \text{ cm}^{-3}$. C–V measurements made at 1 MHz revealed large values for N_{SS} and the interface charge density (2–$3.5 \times 10^{12} \text{ cm}^{-2}\text{·eV}^{-1}$ and $\sim 8 \times 10^{10}$–$10^{12} \text{ cm}^{-2}\text{·eV}^{-1}$, respectively) in the as-grown films. Attributed to dangling Si bonds produced by the ultraviolet radiation, the defects were greatly reduced by annealing the films in N_2 at 250°C for 30 min. The resulting films were measured to have interface charge densities in the $\sim 8 \times 10^{10}$–$2.2 \times 10^{11} \text{ cm}^{-2}$ interval (corresponding to $T_s = 70$ and 100°C, respectively) and interface state densities of $< 5 \times 10^{11} \text{ cm}^{-2}\text{·eV}^{-1}$ to $10^{12} \text{ cm}^{-2}\text{·eV}^{-1}$.

6.3. ALUMINUM OXIDE

Solanki, Ritchie and Collins (194) demonstrated in 1983 that stoichiometric aluminum oxide (Al_2O_3) films could be deposited uniformly over 3-in.-diameter wafers at 200°C by photodissociating mixtures of trimethylaluminum and N_2O with an ArF or KrF laser beam. At a source wavelength of 248 nm and an average laser power of 10 W, film deposition rates up to 2000 Å·min^{-1} were measured at the substrate temperature of 350°C. While higher deposition rates were observed at 193 nm, it was necessary to reduce the TMA partial pressure from 80–120 to 30 mTorr in order to obtain films of uniform thickness. The carbon content in films (deposited at 350°C) was <1 atom %, and the pinhole defect density was measured to be $<0.2/\text{cm}^2$. All of these experiments were carried out in parallel geometry, and Deutsch et al. (390), subsequently demonstrated that irradiating the growing Al_2O_3 film with 193- or 248-nm fluences as low as 1 mJ·cm^{-2} measurably increased the film's refractive index. An increase in the index from ~ 1.6 to 1.7, for example, was observed when the surface was illuminated with 1.6 mJ·cm^{-2} of 193-nm radiation. Ishida et al. (391) photodeposited Al_2O_3 layers on Si in a UHV system and showed that carbon-free dielectric films could be deposited at room temperature by photodissociating N_2O *after* TMA had already decomposed at the surface.

Hudson and co-workers (392) synthesized several novel precursors from the aluminum β-diketonate family as a means of photodepositing Al_2O_3. Containing *both* the necessary aluminum and oxygen atoms, these molecules absorb strongly over a broad spectral region in the UV, making them suitable for photodissociation by a lamp. One of the diisopropoxyaluminum diketonates $[Al(OPr^i)_2Acac(OMe)]$, which contains CH_3 and OCH_3 ligands, was synthesized and studied as a precursor. Although its vapor pressure at 120°C is only ~ 0.03 Torr, stoichiometric Al_2O_3 films having a carbon content < 1 atom % were deposited with a growth rate of $0.25 \, \mu m \cdot h^{-1}$.

6.4. OTHER OXIDES AND NITRIDES

Several other oxides have been deposited by photo-CVD. Matsui and co-workers (125) successfully deposited TaO_x films from mixtures of $TaCl_5$ and O_2. At 500°C, the growth rate was determined to be $> 30 \, Å \cdot min^{-1}$ for a total reactor pressure of 1 Torr. Also, TaO_x deposition rates are inversely proportional to growth temperature but, without the external UV radiation, the deposition rate falls to $2 \, Å \cdot min^{-1}$ at 400°C. At 350°C, Yamagishi and Tarui (126) deposited amorphous TaO_x ($x \approx 0.2$) films at a rate of $120 \, Å \cdot min^{-1}$ by photodissociating pentamethoxytantalum $[Ta(OCH_3)_5]$ in the presence of O_2. X-ray analysis revealed that the non-stoichiometric films were composed of TaO_2 and Ta_2O_5.

Tin dioxide films 1000 Å thick have been deposited from mixtures of $SnCl_4$ and N_2O vapor. Kunz et al. (393) irradiated fused silica substrates at room temperature and normal incidence in the presence of such a mixture, and for laser energy fluences of 10–20 mJ/cm² highly conductive SnO_2 films resulted. Resistivities as low as $4 \times 10^{-2} \, \Omega \cdot cm$ were measured. At lower fluences, $SnOCl_2$ is formed (presumably because modest surface heating by the laser is conducive to the desorption of Cl) and the film resistivity rises rapidly. For fluences beyond 70 mJ/cm², in contrast, ablation becomes dominant and the net deposition rate vanishes. Lines of 10-μm width were also deposited by proximity printing through a mask.

Arnone et al. (394) have demonstrated that films and single crystals of chromium oxides can be deposited on a variety of substrates by photo-lyzing chromyl chloride, CrO_2Cl_2. Upon absorbing a single photon in the 488–515 nm region ($\hbar\omega \simeq 2.5 \, eV$), CrO_2Cl_2 molecules adsorbed onto the substrate are photodissociated, apparently by the two-step sequence:

$$CrO_2Cl_2 \longrightarrow CrO_2Cl + Cl \longrightarrow CrO_2 + Cl + Cl$$

yielding chromium oxide and free Cl atoms. While the deposition process is initiated photochemically, thermal processes rapidly become important as the amount of laser power absorbed by the growing film rises beyond tens of milliwatts. The composition of lines written on Si(100) substrates at scanning speeds of $20 \, \mu m \cdot s^{-1}$ was a mixture of Cr_2O_3 and the ferromagnetic oxide CrO_2. Single crystals, up to almost 1 mm in length, were also grown at rates up to $3 \, \mu m \cdot s^{-1}$ by pyrolytic processes.

Although not strictly thin films, the vanadium oxide catalysts produced by Anpo and co-workers (395) should be mentioned. In the presence of butene vapor (several Torr), $VOCl_3$ was irradiated with the UV from a high-pressure Hg lamp. Reaction of adsorbed $VOCl_3$ molecules with OH groups at the surface of the Vycor glass substrate resulted in VO_x formation. Also, the oxides of both tin and iron have been photodeposited (396) onto porous Vycor glass by first impregnating the glass with $(CH_3)_3SnI$ or $Fe(CO)_5$. After irradiation with 254-, 266-, or 310-nm photons, the glass is heated to 650°C to remove the volatile photoproducts and, at the same time, produce metal oxide. Analysis of the result shows that the glass consolidates around the iron oxide particles but not around the tin oxide. Consequently, photo-chemically generating tin oxide on the surface of porous glass appears to be an approach for maintaining porosity of the glass in a spatially selective manner.

Reduction of the growth temperature of TiN films and 35–300% improvements in the film deposition rate were realized by Motojima and Mizutani (397) by irradiating $TiCl_4$, NH_3 (or N_2), and H_2 gas mixtures with a deuterium lamp. In the absence of the UV radiation, and using N_2 as the nitrogen donor, substrate temperatures above 800°C were required in order to observe film growth. With the lamp, however, film deposition occurred at 700°C and, at 900°C, the deposition rate, $2.1 \, \mu m \cdot h^{-1}$, was a factor of 3 larger than that observed without the UV source. Substituting NH_3 for N_2 as the N precursor enables film deposition to occur at temperatures as low as 450°C, and at 800°C the TiN deposition rate ($55 \, \mu m \cdot h^{-1}$) is $\sim 35\%$ higher than that measured without the lamp radiation. Consequently, the addition of the D_2 lamp to the TiN reactor permitted deposition temperatures to be reduced 50–100°C below those normally required.

The performance of InP MISFETs is critically dependent upon the processing temperature since phosphorus depletion (owing to the large vapor pressure of phosphorus) degrades the InP surface. It is necessary, therefore, to pursue deposition techniques suitable for fabricating InP gate insulators at low temperature. Phosphorus nitride (P_3N_5) has several attractive properties as a dielectric for this device, and Jeong et al. (398) recently demonstrated that Au/n-InP Schottky diodes fabricated from P_3N_5 layers photodeposited onto InP have enhanced Schottky barrier heights. Thin (50-Å) P_3N_5 films were deposited on n-InP at 250°C by photolyzing a mixture of PCl_3, NH_3, and H_2 at 185 nm with a lamp. The deposition rate was measured to be

$12.5\,\text{Å·min}^{-1}$, and the refractive index of the film was 1.9. Schottky diodes incorporating the thin P_3N_5/n-InP films exhibited reverse leakage currents as low as 0.1 nA at 1-V bias with an ideality factor of 1.08 and a breakdown voltage of 30 V. The Schottky barrier height for the diode was 0.81 eV.

6.5. STATUS OF DIELECTRIC PHOTODEPOSITION

The electrical and structural properties of photodeposited SiO_2, Si_3N_4, and Al_2O_3 dielectrics have not only distinguished this class of films as examples of the low-temperature-processing potential of photochemical vapor deposition but have made them competitive with films grown by conventional thermal or plasma processes. This is particularly true for silicon dioxide, which exhibits, with proper surface treatment, excellent interfacial state densities, breakdown voltages, and low defect densities. Step coverage is conformal, and the film refractive index is controllable through the silane and oxygen precursor partial pressures. Moreover, these results were obtained with low-pressure Hg lamps, which are attractive for use in an industrial environment since they are cost effective and reliable. For Si_3N_4 films, the substrate temperature can be reduced by several hundred degrees Celsius for photodeposition while still obtaining etch rates and electrical characteristics that are comparable to films grown thermally. Photo-CVD aluminum oxide films have consistently low carbon content despite the use of Al-alkyl precursors. Because of these promising results and the success of depositing an array of other insulators such as P_3N_5 that are particularly valuable for III–V FETs (field effect transistors) sensitive to processing temperature, continued development of photo-CVD dielectrics is warranted and the broader application of their properties to microelectronics, in particular, is likely.

CHAPTER

7

MISCELLANEOUS MATERIALS
AND APPLICATIONS

Photochemical vapor deposition is a process of tremendous flexibility and, not surprisingly, has also been applied toward the deposition of materials that are not of direct interest to the microelectronics industry, for example. This chapter briefly describes several materials and applications that have benefited from one or more of the unique characteristics of photo-CVD. As the field matures and new precursors develop, the scope of the materials accessible will undoubtedly widen, but these films are representative of the broad applicability of photo-CVD (see Table 19).

Substrate temperature is an important processing parameter for many electronic and photonic materials, and the recently developed high-T_c superconductors are in this category. Metal oxide films that are related to the Bi–Sr–Ca–Cu–O superconductor system have been deposited by photo-CVD (399) at temperatures between 300 and 600°C. Oxides of Bi, Sr, Ca, Cu, and Ca–Cu were grown on MgO(100) substrates by photodissociating mixtures of the appropriate metal alkyl and either O_2, N_2O, or NO_2 with a low-pressure Hg lamp. For Bi, the precursor was $Bi(C_6H_5)_3$, whereas the other metal oxides involved the alkyl $M(DPM)_2$, where M is the metal atom (Sr, Ca, Cu) and DPM is the dipyvaloymethanate group. At 300°C, 1000-Å-thick films of α-Bi_2O_3 were deposited in 2 h, but in the absence of surface irradiation no deposition occurred. Photo-CVD of $SrCO_3$ films yielded (001)-oriented films at 400°C, whereas amorphous films were deposited at the same temperature when the lamp was turned off. Similar improvements in growth rate or crystallinity were observed for $CaCO_3$, CuO, and Ca_2CuO_3 films deposited at 400°C when UV photons were present at the surface.

Improving the corrosion and wear characteristics of metals with thin films deposited by numerous techniques has been studied extensively, but Fotakis and co-workers have demonstrated (400) the flexibility of photo-CVD in coating Al- or Ni-based alloys or modifying their surfaces. Boron and hafnium have been deposited on various metal substrates by photodissociating BCl_3 at 193 nm and HfI_4 at 248 nm, respectively. The use of ArF (193-nm) laser photons rather than 248-nm radiation improved the surface morphology of

Table 19. Miscellaneous Materials Deposited by Photo-CVD

Material	Precursor(s)	Optical Source	λ (nm)	Deposition Conditions	Refs.	Results
Bi_2O_3	$Bi(C_6H_5)_3$ and N_2O, NO_2, or O_2	Hg lamp	185, 254	MgO(100) substrate temperature: 300°C; $Bi(C_6H_5)_3$ precursor heated to 120°C and entrained in 50 sccm of Ar; 110 W Hg lamp	399	1000 Å thick film was deposited in 2 h; polycrystalline α-Bi_2O_3.
CaO, $CaCO_3$	$Ca(DPM)_2$ and N_2O, NO_2, or O_2 (DPM = dipyvaloylme-thanate)	Hg lamp	185, 254	$CaCO_3$: 400°C; CaO: 400–500°C; $Ca(DPM)_2$ heated to 220°C; MgO substrate	399	N_2O suppresses $CaCO_3$ deposition at 400°C and favors CaO growth; $CaCO_3$ grains in films deposited at 400°C are less than 0.2 μm in diameter.
CuO	$Cu(DPM)_2$ and N_2O, NO_2, or O_2	Hg lamp	185, 254	$Cu(DPM)_2$ heated to 130°C; MgO(100) substrate; substrate temperatures: 300 and 400°C	399	8000 Å deposited in 1 h with 5.7 Torr O_2; polycrystalline films–UV radiation improves film crystallinity.
SrO, $SrCO_3$	$Sr(DPM)_2$ and N_2O, NO_2, or O_2	Hg lamp	185, 254	$Sr(DPM)_2$ heated to 200°C; MgO substrate; 400–600°C substrate temperature	399	SrO film deposited at 600°C; at 500–550°C, polycrystalline $SrCO_3$ was grown–(001) orientation is improved by using NO_2 rather than O_2 precursor.

164

Material	Precursor	Light source	Wavelength (nm)	Conditions	Ref.	Comments
Polymers	Tetrafluoroethylene	Hg lamp	185, 254	Gold-coated quartz substrates; precursor pressure: 0.5–10.0 Torr; 1000 W lamp source	405	Thin (< 1000 Å thick) polymer films were deposited on gold-coated quartz substrates; threshold wavelength was determined to be 215 nm; analysis of the gas phase by-products revealed fluorocarbons as large as C_9F_{18}. Polymerization occurred at the substrate surface.
Organics— siloxanes and phtha-lates	Tetramethyltetra-phenyltrisiloxane [DC-704], bis(2-ethylhexyl)-phthalate (DEHP)	Kr or Xe microwave-excited resonance lamps	125–175	300 Torr lamps driven at 2.45 GHz; substrate temperature: 33°C; aluminum, platinum, or gold substrates (microbalance electrodes)	404	Deposition of siloxane and phthalate organic films was demonstrated by photodissociation in the VUV with rare gas lamps; photodeposition required wavelengths $\leqslant 200$ nm.
TiB_2	$TiCl_4/BCl_3/H_2/Ar$	D_2 lamp		850°C Cu substrate temperature; $TiCl_4$ $+ BCl_3$ mass flow rate: 0.4 sccm	401	For B/Ti flow ratio of 1:7 and with the lamp on, B/Ti ratio in the film was 2.20–2.26; with no UV radiation and B/Ti flow ratio varied from 1 to 15, B/Ti composition in the films ranged from 1.70 to 2.35; lamp lowers required deposition temperature by 50°C.

boron films on Al alloys, and the deposition of Hf films on Ni alloys serves to improve their anticorrosion properties.

Motojima and Mizutani (401), have deposited several transition metal carbides, nitrides, and borides with the aid of a deuterium lamp. Titanium boride (TiB_2) films were deposited on copper, for example, from $TiCl_4/BCl_3/H_2/Ar$ mixtures at 850°C. With lamp irradiation of the surface, film growth rates increased by a factor of 1.5–2.5 as compared to conventional (thermal) CVD and the growth temperature was reduced by 50°C. Film morphology was improved in the presence of surface photons, suggesting that the VUV radiation from the D_2 lamp promotes surface reactions. With an ArF excimer laser, Elders et al. (402) obtained TiB_2 growth rate enhancements larger than 40%. Titanium diboride is a refractory material that has several excellent characteristics as a coating for resisting corrosion and minimizing tool wear.

In an effort to develop a passivating layer for GaAs, Mackey and co-workers (403) have demonstrated the photo-CVD growth of CaF_2 on GaAs at temperatures as low as 100°C. Polycrystalline films several thousand angstroms thick were grown by photolyzing bis(1,1,1,5,5,5-hexafluoro-2,4-pentanedionato)calcium (II) [$Ca(Hfacac)_2$] with the radiation from a high pressure xenon arc lamp. The absorption spectrum of the acetyl acetonate precursor peaks near 260 nm, has a spectral breadth of 34 nm (FWHM) and is, therefore, well suited for other optical sources such as the Hg lamp or the KrF laser. Because CaF_2 is transparent deep into the VUV, this photochemical process also holds intriguing possibilities for coating optical components such as mirrors and lenses.

Organic (404) and polymer (405) films such as the siloxanes (404) have also been deposited photochemically. Polymer thins ($< 0.1\,\mu m$ thick) were deposited by photodissociating tetrafluoroethylene with UV radiation from an Hg lamp. Maylotte and Wright (405) showed that the photodeposition process requires photons of $\lambda < 215$ nm, and gas phase by-products as large as C_9F_{18} were identified. Polymerization occurs at the substrate surface. Also, thin organic films were deposited at 37°C on aluminum, gold, or platinum substrates by photodissociating siloxane or phthalate precursors. Optical source wavelengths below 200 nm were found to be necessary for film deposition to occur.

An interesting application of photo-CVD is the fabrication of multilayer X-ray mirrors by Suzuki (406). By combining an ArF laser, which served to photodissociate WF_6 and C_2H_2 precursors in the gas phase, with surface photons provided by an Hg lamp, mirrors consisting of eight pairs of tungsten and carbon layers were produced. For individual layer thicknesses of 50 Å, the reflectivity of the structure was measured to be 40% at the Cu-Kα wavelength of 1.54 Å. All of the films were deposited at 200°C on Si(100) substrates. Despite the fact that the reflectivities attained are only one-half of the theoretical limit (owing to interfacial irregularities and small β phase W crystals dispersed throughout the tungsten layer), photo-CVD is an attractive approach for adapting microfabrication techniques to producing short-wavelength optics because of its low processing temperature and the ability to accurately control individual layer thicknesses with laser intensity.

Another X-ray related application is the use of photo-CVD to repair so-called clear defects in X-ray lithographic masks. Clear defects are those regions of the mask where X-ray absorbing material is unintentionally absent. It is necessary that processes be developed for inserting the missing features and doing so with a spatial resolution of $\leqslant 0.1 \,\mu$m. Putzar, Petzold, and Staiger (407) have demonstrated that the photodeposition of tin from mixtures of $Sn(CH_3)_4$ and O_2 with a frequency-doubled Ar^+ laser yielded deposition rates as high as $800 \,\text{Å·s}^{-1}$ with feature sizes equal to the optical resolution. Also, they noted that "the X-ray opacity of the deposits was comparable to bulk tin".

The growing influence of polymers and ceramics on electronic devices and packaging is placing increasingly stringent demands on existing metallization technology, particularly insofar as the processing temperature and spatial resolution are concerned. Photochemical techniques appear promising in meeting several of the most pressing process requirements and, as an example, Bauer et al. (408) have demonstrated the ability of laser-assisted deposition to pattern metal deposits on a variety of ceramics and polymers. Although not a gas phase process, it nevertheless demonstrates the versatility of photodeposition. The process consists of spraying, spin-coating, or dipping the substrate with a solution of palladium acetate and chloroform. The surface is then exposed to laser radiation through a mask and the residual metal acetate precursor is removed by rinsing the substrate in chloroform. In an electroless solution, palladium deposits serve to catalyze the deposition of copper, yielding metal line widths in the $10–100 \,\mu$m range (depending on the substrate material). The temperature of this process is compatible with polyimide and other low-temperature materials.

CHAPTER

8

SURFACE PROCESSING

It would be remiss to not mention several benefits associated with optical radiation impinging on a surface prior to or during film deposition. While these are not specifically photochemical deposition processes in themselves, several play a role in photo-CVD and have been alluded to earlier. That is, deposition is an overall label given to an interacting set of processes, but several members of that set can be readily identified as being useful in other circumstances. A number of beneficial effects of visible and UV photons on a surface, including cleaning and improved film adhesion, are discussed here.

8.1. OXIDATION AND NITRIDATION

The oxidation and nitridation of silicon (and other semiconductor materials) is crucial to the fabrication of microelectronic devices, and it has long been known that the presence of visible or UV radiation accelerates the process (21). One cause of this effect has been traced to the photogeneration of charge carriers at the surface and emission of electrons into the film that facilitate the decomposition of the oxygen or nitrogen precursor (409–411), but the role of photochemical production of radicals in the gas phase or on the surface must also be acknowledged.

As an example of this process consider the oxidation of Si, which traditionally has entailed reacting dry O_2 with the surface at temperatures beyond 900°C. A considerable reduction in the process temperature can be realized by photodissociating N_2O with the 185-nm photons from a low-pressure lamp. Ishikawa and co-workers (412) have demonstrated that, as illustrated in Figure 74, the enhancement in oxidation rate that is observed with surface illumination is significant for temperatures below 600°C, and in the 300–500°C range roughly 40 Å of oxide is grown in 1/2 hour. Not only is the process temperature lowered, but MOS diodes fabricated from oxides grown at 500°C display no noticeable hysteresis in their $C-V$ characteristics. The fixed charge density at the Si/SiO_2 interface was estimated to be 2×10^{11} cm^{-2}, and the breakdown voltage of the dielectric was measured to be 1.8×10^7 V/cm. Oxynitrides have also been grown on Si(100) by irradiating the surface with excimer laser photons when the substrate is

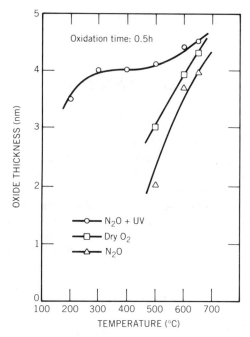

Figure 74. Thickness of SiO$_2$ films grown by thermal oxidation as a function of substrate temperature and for various deposition conditions. The combined use of N$_2$O and Hg illumination of the surface gives rise to an oxidation process for which the oxide thickness is essentially independent of temperature in the 300–500°C range [after Ref. *412*, by permission].

immersed in nitrogen or ammonia gas (*21, 413*). Laser fluences at the substrate typically exceed 0.5 J/cm^2, and dielectric breakdown voltages greater than 7×10^6 V/cm can be obtained.

Enhancing the growth of oxide on GaAs has been investigated by Lu et al. (*411*), who showed that deep-UV photons having energies greater than 4.1 ± 0.2 eV (see Figure 75) increase the oxide thickness by more than a factor of 5. All of the evidence points to the photoemission of electrons from the *n*-type GaAs substrate into the growing oxide as the responsible mechanism. Electrons arriving in the oxide produce oxygen anions (presumably O$_2^-$) that are transported under the influence of an internal electric field to the GaAs/oxide interface. Longer wavelength (i.e., visible and near-UV) photons have the effect of increasing the sticking probability of oxygen to GaAs by 3 orders of magnitude (*414*).

A related effect is the improvement of the adhesion or growth rate of metal (*415–417*) or semiconductor films (*418, 419*). The latter stems from the production of carriers at semiconductor surfaces by above-bandgap radiation,

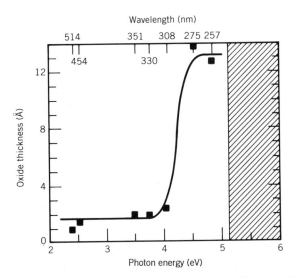

Figure 75. Enhancement in the growth of oxide on GaAs by irradiation of the surface with photons having energies greater than 4.1 eV [after Ref. *411*].

and the former is attributable to several interactions, including electronic excitation at the interface (*415*).

8.2. SURFACE CLEANING

The removal of oxygen, hydrocarbons, and metallic contaminants from semiconductor surfaces by irradiating the substrate with UV photons has been shown to be an effective and relatively straightforward process (*420–423*). Both Si and GaAs surfaces have been studied extensively. By illuminating the surface with a low-pressure Hg lamp while flowing H_2, Si(100) substrates have been cleaned (*421*) at 800°C as thoroughly as by conventional methods but by a process that is simpler to implement. Synchrotron radiation is also effective in breaking Si—H bonds on HF-passivated Si(111) substrates (*422*), thereby cleaning the surface without the need to resort to higher processing temperatures.

Surface illumination from a low pressure Hg lamp ($\lambda = 254$ nm), combined with ozone photochemically produced by the 185-nm photons also generated by the lamp, has been shown to be capable of removing C and O from GaAs surfaces prior to the growth of films by MBE or MOMBE (*423*). As shown in Figure 76, UV/ozone cleaning of GaAs lowers oxygen and carbon

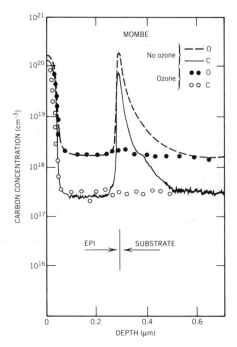

Figure 76. SIMS depth profiles of carbon and oxygen concentrations in 0.3-μm-thick GaAs films grown on GaAs by metal-organic molecular beam epitaxy (MOMBE). Treating the surface with UV radiation (254 nm) and UV-produced ozone lowers the C and O concentrations to below the detectability limit of SIMS ($\sim 4 \times 10^{17}$ cm^{-3}) [after Pearton et al. (*423*), by permission].

concentrations at the interface of 0.3-μm GaAs epilayers grown on GaAs by MOMBE to levels below the detectability limit of SIMS [roughly 4×10^{17} cm^{-3} in Ref. (*407*)]. This cleaning technique has much less effect on MOCVD-grown interfaces, apparently because atomic hydrogen produced in the MOCVD growth process is effective in removing contaminants without the need for further treatment. Static charge buildup on the surface of dielectrics can also be suppressed with a low-pressure Hg lamp (*424*).

CHAPTER

9

FUTURE PROSPECTS AND CONCLUSIONS

Despite the rapid expansion in the field of photochemical vapor deposition in the last decade, it is still clearly in the early stages of development. The attainable film quality and the extent of the materials that can be deposited by this approach, for example, have only been addressed in the recent past. Because of the compatibility of photo-CVD with other optical, plasma, and ion beam *in situ* processing techniques and its demonstrated ability to produce high-quality damage-free films at reduced temperatures, increased scientific and engineering effort into this new deposition technique is both warranted and inevitable. In a broader sense, the capability of assisting or controlling surface and gas phase reactions that light affords is too versatile to not be exploited.

Several themes have recurred throughout this volume. For most of the films examined to date, photo-CVD (1) increases growth rates; (2) permits film growth at reduced temperatures, as compared to thermal deposition (CVD); and (3) results in improved surface morphologies. Each of these is the culmination of the complex interactions of photons and optically produced atoms and molecular radicals with the surface. Several processes such as photo-stimulated desorption of products, photodissociation of adsorbed species, and the photogeneration of carriers have been shown to be instrumental in bringing about such desirable film growth characteristics. However, much effort lies ahead in tailoring the gas phase and adlayer photochemistry so as to (1) optimally transfer energy from the optical source to the precursor and (2) excite the precursors in such a way as to discourage the formation of undersirable products. Carbon incorporation into metal films deposited from $Al(CH_3)_3$ at wavelengths below $\sim 220\,nm$ is a prominent example of the latter, and the introduction of precursors such as dimethyl-aluminum hydride or its halide counterparts represents an initial step in alleviating that difficulty. More detailed spectroscopic and kinetics studies of the photofragments produced from conventional precursors will lead to new laser(or lamp)/precursor combinations, as well as increased activity in the synthesis of novel precursors such as the metal β-diketonates reported by Hudson et al. (*392*), the AlAs adduct prepared by Zinck et al. (*354*) and the calcium acac used by Mackey and co-workers (*403*). Precursor design also must consider optimizing the adsorption of the precursor itself or a critical photofragment onto the substrate, as well as the desorption of ligands or other surface reaction products. In short, the obstacle to realizing the full potential of photodeposition is the continuing need to elucidate the detailed physics and chemistry occurring at surfaces that will allow one to enhance particular photoprocesses and generate specific film compositions and structures.

In the near future, the optical production of transient species immediately in the vicinity of the substrate with lasers will undoubtedly be increasingly pursued. Producing chemically active reactants only as they are needed from relatively nonreactive feedstock gases of low toxicity is attractive in that it minimizes the exposure of the reactor itself to toxic and often corrosive gases but, more importantly, minimizes their handling by operating staff.

However, for broad-area deposition applications, emphasis will likely continue to rest on lamps as optical sources. The development of high average power, microwave-excited, rare gas–halide lamps (as discussed in Chapter 3) offers considerably higher duty cycle operation than that available from pulsed lasers and at lower cost. For multistep *in situ* processing, it may become necessary to introduce two or more optical sources, each tailored to perform a specific photochemical function. As Kachi et al. (337) have demonstrated for GaAs, one source may be directed toward a gas phase process and the second designed for exciting surface reactions. The latter would be beneficial for improving and controlling the crystalline structure and electrical properties of semiconductor films. Intentionally combining photochemical and pyrolytic processes is also often beneficial as the photolytic process might act as the rate-limiting mechanism, but heating the surface also provides the necessary mobility for adatoms or encourages desorption of contaminants.

New materials remain to be explored. Photoepitaxy of the wider band gap III–V binary and ternary compounds such as AlGaN is of interest, and efforts devoted to controlling the electrical properties of II–VI films, in particular, by photo-activated substitutional doping are needed. With the recent realization of II–VI compound semiconductor junction lasers operating in the blue/green, both subjects have became of greater interest. The high-T_c superconductors (Y–Ba–Cu–O) are attractive candidates for photo-CVD since ion beam (425) and thermal (pyrolytic) (426) decomposition of metal organics has proven successful for the patterned deposition of these materials and the constituent oxides of the Bi–Sr–Ca–Cu–O system have been photodeposited by Koinuma et al. (399). Other materials that appear to be well suited for photodeposition techniques are ceramic and diamond films.

One theme of this review is that gas phase and surface photochemical interactions resulting in film deposition are no longer of intellectual interest only but are of commercial utility as well, and it appears that this trend will accelerate in the foreseeable future.

REFERENCES

1. H. Romeyn, Jr., and W. A. Noyes, Jr., *J. Am. Chem. Soc.* **54**, 4143 (1932).
2. H. J. Emeléus and K. Stewart, *Trans. Faraday Soc.* **32**, 1577 (1936).
3. H. Niki and G. J. Mains, *J. Phys. Chem.* **68**, 304 (1964).
4. C. H. van der Brekel and P. J. Severin, *J. Electrochem. Soc.* **119**, 372 (1972).
5. J. L. Vossen and W. Kern, eds., *Thin Film Processes II.* Academic Press, San Diego, CA, 1991.
6. T. Inushima, N. Hirose, K. Urata, K. Ito, and S. Yamazaki, *Appl. Phys. A* **47**, 229 (1988).
7. S. C. Su, *Solid State Technol.* **24**, 72–82 (1981).
8. L. P. Welsh, J. A. Tuchman, and I. P. Herman, *J. Appl. Phys.* **64**, 6274 (1988).
9. See, for example, Japanese Patents 59,194,452, "Multilayer wiring for integrated circuits" (Mitsubishi Electric Co., awarded Nov. 5, 1984) and 59,218,722, "Photochemical film deposition for integrated circuit fabrication" (Ushio, Inc., awarded Dec. 10, 1984).
10. K. Yano, S. Tanimura, T. Ueda, and T. Fujita, *1987 Symposium on VLSI Technology*, p. 71. Business Center for Academic Society of Japan, Karuizawa, Japan, 1987.
11. M. Akazawa, H. Hasegawa, and E. Ohue, *Jpn. J. Appl. Phys.* **28**, L2095 (1989).
12. K. Hirabayashi and H. Kozawaguchi, *Jpn. J. Appl. Phys.* **28**, 814 (1989).
13. R. M. Osgood, Jr., *Annu. Rev. Phys. Chem.* **34**, 77 (1983).
14. R. M. Osgood, Jr., and H. H. Gilgen, *Annu. Rev. Mater. Sci.* **15**, 549 (1985).
15. R. M. Osgood, Jr., and T. F. Deutsch, *Science* **227**, 709 (1985).
16. Y. Rytz-Froidevaux, R. P. Salathé, and H. H. Gilgen, *Appl. Phys. A* **37**, 121 (1985).
17. F. A. Houle, *Appl. Phys. A* **41**, 315 (1986).
18. J. Haigh and M. R. Aylett, *Prog. Quantum Electron.* **12**, 1 (1988).
19. I. P. Herman, *Chem. Rev.* **89**, 1323 (1989).
20. D. Bäuerle, *Chemical Processing with Lasers*, Springer Ser. Mater. Sci., Vol. 1. Springer-Verlag, Berlin, 1986.
21. I. W. Boyd, *Laser Processing of Thin Films and Microstructures*, Springer Ser. Mater. Sci., Vol. 3. Springer-Verlag, Berlin, 1987.
22. D. J. Ehrlich and J. Y. Tsao, eds., *Laser Microfabrication: Thin Film Processes and Lithography.* Academic Press, San Diego, CA, 1989.

23. K. G. Ibbs and R. M. Osgood, Jr., eds., *Laser Chemical Processing for Micro-electronics.* Cambridge Univ. Press, London, 1989.

24. V. R. McCrary and V. M. Donnelly, in *Chemical Vapor Deposition* (K. F. Jensen, ed.). Academic Press, San Diego, CA (in press).

25. R. L. Abber, in *Handbook of Thin Film Deposition Processes and Techniques,* (K. K. Schuegraf, ed.), pp. 270–290. Noyes Data Corp., Park Ridge, NJ, 1988.

26. Y. Aoyagi, T. Meguro, and S. Iwai, in *Low Temperature Epitaxial Growth of Semiconductors* (T. Hariu, ed.), pp. 199–290. World Scientific, Singapore, 1991.

27. D. J. Ehrlich, R. M. Osgood, Jr., and T. F. Deutsch, *J. Vac. Sci. Technol.* **21**, 23 (1982).

28. F. A. Houle and S. D. Allen, *Laser Deposition and Etching,* course notes. American Vacuum Society Education Committee.

29. H. Okabe, *Photochemistry of Small Molecules.* Wiley, New York, 1978.

30. J. G. Calvert and J. N. Pitts, Jr., *Photochemistry.* Wiley, New York, 1966.

31. W. A. Noyes and P. A. Leighton, *The Photochemistry of Gases.* Dover, New York, 1966.

32. U. Itoh, Y. Toyoshima, H. Onuki, N. Washida, and T. Ibuki, *J. Chem. Phys.* **85**, 4867 (1986).

33. M. J. Mitchell, X. Wang, C. T. Chin, M. Suto, and L. C. Lee, *J. Phys. B* **20**, 5451 (1987).

34. M. Suto, C. Ye, and L. C. Lee, *J. Chem. Phys.* **89**, 160 (1988).

35. S. P. Zuhoski, K. P. Killeen, and R. M. Biefeld, *Mater. Res. Soc. Symp. Proc.* **101**, 313 (1988).

36. G. A. Hebner, K. P. Killeen, and R. M. Biefeld, *J. Cryst. Growth* **98**, 293 (1989).

37. J. J. Zinck, P. D. Brewer, J. E. Jensen, G. L. Olson, and L. W. Tutt, *Mater. Res. Soc. Symp. Proc.* **101**, 319 (1988).

38. Y. Fujita, S. Fujii, and T. Iuchi, *J. Vac. Sci. Technol. A* **7**, 276 (1989).

39. J. Heidberg, R. Daghighi-Ruhi, H. von Weyssenhoff, and A. Habekost, *Mater. Res. Soc. Symp. Proc.* **101**, 221 (1988).

40. M. Rothschild, in *Laser Microfabrication: Thin Film Processes and Lithography* (D. J. Ehrlich and J. Y. Tsao, eds.), pp. 163–230, and references cited therein. Academic Press, San Diego, CA, 1989.

41. T. Beuermann and M. Stuke, *Appl. Phys. B* **49**, 145 (1989).

42. Y. Zhang and M. Stuke, *Jpn. J. Appl. Phys.* **27**, L1349 (1988).

43. Y. Zhang, T. Beuermann, and M. Stuke, *Appl. Phys. B* **48**, 97 (1989).

44. G. S. Higashi and M. L. Steigerwald, *Appl. Phys. Lett.* **54**, 81 (1989).

45. J. L. Brum, S. Deshmukh, and B. Koplitz, *J. Chem. Phys.* **93**, 7946 (1990).

46. R. D. Kenner, F. Rohrer, and F. Stuhl, *Chem. Phys. Lett.* **116**, 374 (1985).

47. K. Watanabe, *J. Chem. Phys.* **22**, 1564 (1964).

48. D. H. Lowndes, D. B. Geohegan, D. Eres, S. J. Pennycook, D. N. Mashburn, and G. E. Jellison, Jr., *Appl. Phys. Lett.* **52**, 1868 (1988).

49. C. L. Sam and J. T. Yardley, *J. Chem. Phys.* **69**, 4621 (1978).

50. V. M. Donnelly, A. P. Baronavsky, and J. P. McDonald, *Chem. Phys.* **43**, 271 (1979).

51. K. E. Lewis, D. M. Golden, and G. P. Smith, *J. Am. Chem. Soc.* **106**, 3905 (1984).

52. W. H. Breckenridge and N. Sinai, *J. Phys. Chem.* **85**, 3557 (1981).

53. W. H. Breckenridge and G. M. Stewart, *J. Am. Chem. Soc.* **108**, 364 (1986).

54. T. A. Seder, S. P. Church, A. J. Ouderkirk, and E. Weitz, *J. Am. Chem. Soc.* **107**, 1432 (1985).

55. E. Weitz, *J. Phys. Chem.* **91**, 3945 (1987), and references cited therein.

56. For a recent review of Group VI carbonyl photolysis, see Y. Ishikawa, C. E. Brown, P. E. Hackett, and D. M. Rayner, *J. Phys. Chem.* **94**, 2404 (1990).

57. T. R. Fletcher and R. N. Rosenfeld, *J. Am. Chem. Soc.* **107**, 2203 (1985).

58. W. Tumas, B. Gitlin, A. M. Rosan, and J. T. Yardley, *J. Am. Chem. Soc.* **104**, 55 (1982).

59. T. A. Seder, S. P. Church, and E. Weitz, *J. Am. Chem. Soc.* **108**, 4721 (1986).

60. G. W. Tyndall and R. L. Jackson, *J. Am. Chem. Soc.* **109**, 582 (1987).

61. W. Radloff, H. Hohmann, H.-H. Ritze, and R. Paul, *Appl. Phys. B* **49**, 301 (1989).

62. H. Ohashi, K. Inoue, Y. Saito, A. Yoshida, H. Ogawa, and K. Shobatake, *Appl. Phys. Lett.* **55**, 1644 (1989).

63. K. Kameta, M. Ukai, R. Chiba, K. Nagano, N. Kouchi, Y. Hatano, and K. Tanaka, *J. Chem. Phys.* **95**, 1456 (1991).

64. A. Terenin, *Phys. Rev.* **36**, 147 (1930).

65. A. Terenin and B. Popov, *Phys. Z. Sowjetunion* **2**, 299 (1932).

66. D. B. Geohegan and J. G. Eden, *J. Chem. Phys.* **81**, 5336 (1984).

67. J. Berkowitz and W. A. Chupka, *J. Chem. Phys.* **45**, 1287 (1966).

68. D. B. Geohegan and J. G. Eden, *Appl. Phys. Lett.* **45**, 1146 (1984).

69. A. R. Calloway, T. A. Galantowicz, and W. R. Fenner, *J. Vac. Sci. Technol. A* **1**, 534 (1983); A. R. Calloway and W. R. Fenner, *Aerosp. Corp. Rep.* **ATR-84(8503)-1** (1984).

70. K. K. King, Ph.D. thesis, University of Illinois, Urbana, 1989.

71. R. Burnham, *Appl. Phys. Lett.* **30**, 133 (1977).

72. B. Koplitz, Z. Xu, and C. Wittig, *Appl. Phys. Lett.* **52**, 860 (1988).

73. H. H. Gilgen, C. J. Chen, R. Krchnavek, and R. M. Osgood, Jr., in *Laser Processing and Diagnostics* (D. Bäuerle, ed.), p. 225. Springer-Verlag, Berlin, 1984.

74. N. Suzuki, C. Anayama, K. Masu, K. Tsubouchi, and N. Mikoshiba, *Jpn. J. Appl. Phys.* **25**, 1236 (1986).

75. H. Itoh, M. Watanabe, S. Mukai, and H. Yajima, *J. Cryst. Growth* **93**, 165 (1988).

76. H. Okabe, M. K. Emadi-Babaki, and V. R. McCrary, *J. Appl. Phys.* **69**, 1730 (1991).

77. M. Fischer, R. Luckerath, P. Balk, and W. Richter, *Chemtronics* **3**, 156 (1988).

78. J. Haigh, *J. Mater. Sci.* **18**, 1072 (1983).

79. E. Tokumitsu, T. Yamada, M. Konagai, and K. Takahashi, *Mater. Res. Soc. Symp. Proc.* **101**, 307 (1988).

80. E. Tokumitsu, T. Yamada, M. Konagai, and K. Takahashi, *J. Vac. Sci. Technol. A* **7**, 706 (1989).

81. T. Cacouris, G. Scelsi, P. Shaw, R. Scarmozzino, R. M. Osgood, and R. R. Krchnavek, *Appl. Phys. Lett.* **52**, 1865 (1988).

82. J. G. Clark and R. G. Anderson, *Appl. Phys. Lett.* **32**, 46 (1978).

83. V. R. McCrary and V. M. Donnelly, *J. Cryst. Growth* **84**, 253 (1987).

84. T. F. Deutsch, D. J. Ehrlich, D. D. Rathman, D. J. Silversmith, and R. M. Osgood, Jr., *Appl. Phys. Lett.* **39**, 825 (1981).

85. M. P. Irion and K. L. Kompa, *J. Chem. Phys.* **76**, 2338 (1982).

86. T. Ibuki, A. Hiraya, K. Shobatake, Y. Matsumi, and M. Kawasaki, *Chem. Phys. Lett.* **160**, 152 (1989).

87. K. G. Ibbs and M. L. Lloyd, *Opt. Laser Technol.* **15**, 35 (1983).

88. J. R. McDonald, A. P. Baronavsky, and V. M. Donnelly, *Chem. Phys.* **33**, 161 (1978).

89. P. G. Wilkinson and R. S. Mulliken, *J. Chem. Phys.* **23**, 1895 (1955).

90. M. Rothschild and D. J. Ehrlich, unpublished (1986).

91. S. J. C. Irvine, J. B. Mullin, D. J. Robbins, and J. L. Glasper, *Mater. Res. Soc. Symp. Proc.* **29**, 253 (1984).

92. C. J. Chen and R. M. Osgood, Jr., *J. Chem. Phys.* **81**, 327 (1984).

93. T. M. Mayer, G. J. Fisanik, and T. S. Eichelberger, *J. Appl. Phys.* **53**, 8462 (1982).

94. D. K. Flynn, J. I. Steinfeld, and D. S. Sethi, *J. Appl. Phys.* **59**, 3914 (1986).

95. P. Hess, *Spectrochim. Acta* **46A**, 489 (1990).

96. R. Halonbrenner, J. R. Huber, U. Wild, and J. J. Gunthard, *J. Phys. Chem.* **73**, 3929 (1968).

97. G. Nathanson, B. Gitlin, A. M. Rosan, and J. T. Yardley, *J. Chem. Phys.* **74**, 361 (1981).

98. D. J. Ehrlich, R. M. Osgood, Jr., and T. F. Deutsch, *J. Electrochem. Soc.* **128**, 2039 (1981).

99. J. T. Yardley, B. Gitlin, G. Nathanson, and A. M. Rosan, *J. Chem. Phys.* **74**, 370 (1981).

100. Y. Rytz-Froidevaux, R. P. Salathé, and H. H. Gilgen, *Mater. Res. Soc. Symp. Proc.* **17**, 29 (1983).

101. J. F. Osmundsen, C. C. Abele, and J. G. Eden, *J. Appl. Phys.* **57**, 2921 (1985).

102. S. L. Baughcum and S. R. Leone, *Chem. Phys. Lett.* **89**, 183 (1982).

103. R. Karlicek, J. A. Long, and V. M. Donnelly, *J. Cryst. Growth* **68**, 123 (1984).

104. V. M. Donnelly, D. Brasen, A. Appelbaum, and M. Geva, *J. Appl. Phys.* **58**, 2022 (1985).

105. K. N. Tanner and A. B. F. Duncan, *J. Am. Chem. Soc.* **73**, 1164 (1951).

106. R. McDiarmid, *J. Chem. Phys.* **61**, 3333 (1974).

107. H. B. Gray and N. A. Beach, *J. Am. Chem. Soc.* **85**, 2922 (1963).

108. N. Rösch, M. Kotzian, H. Jörg, H. Schröder, B. Rager, and S. Metev, *J. Am. Chem. Soc.* **108**, 4238 (1986).

109. M. Zelikoff, K. Watanabe, and E. C. Y. Inn, *J. Chem. Phys.* **21**, 1643 (1953).

110. C. Hubrich and F. Stuhl, *J. Photochem.* **12**, 93 (1980).

111. T. Nakayama, M. Y. Kitamura, and K. Watanabe, *J. Chem. Phys.* **30**, 1180 (1959).

112. A. M. Bass, A. E. Ledford, Jr., and A. H. Laufer, *J. Res. Natl. Bur. Stand.* **80A**, 143 (1976).

113. G. DiStefano, M. Lenzi, A. Margani, A. Mele, and C. N. Xuan, *J. Photochem.* **7**, 335 (1977).

114. T. J. Xia, C. Y. R. Wu, and D. L. Judge, *Phys. Scr.* **41**, 870 (1990).

115. P. A. Leighton and R. Mortensen, *J. Am. Chem. Soc.* **58**, 448 (1936).

116. M. S. Chiu, K. P. Shen, and Y. K. Ku, *Appl. Phys. B* **37**, 63 (1985).

117. L. V. Koplitz, D. K. Shuh, Y.-J. Chen, R. S. Williams, and J. I. Zink, *Appl. Phys. Lett.* **53**, 1705 (1988).

118. Q. Mingxin, R. Monot, and H. van den Bergh, *Sci. Sin. (Engl. Ed.)* **A27**, 531 (1984).

119. R. McDiarmid, *J. Mol. Spectrosc.* **39**, 332 (1971).

120. Y. Harada, J. N. Murrell, and H. H. Sheena, *Chem. Phys. Lett.* **1**, 595 (1968).

121. G. Inoue and M. Suzuki, *Chem. Phys. Lett.* **105**, 641 (1984).

122. J. E. Baggott, H. M. Frey, P. D. Lightfoot, and R. Walsh, *Chem. Phys. Lett.* **125**, 22 (1986).

123. J. Fernandez, G. Lespes, and A. Dargelos, *Chem. Phys.* **111**, 97 (1987).

124. T. Tabuchi, K. Yamagishi, and Y. Tarui, *Jpn. J. Appl. Phys.* **26**, L186 (1987).

125. M. Matsui, S. Oka, K. Yamagishi, K. Kuroiwa, and Y. Tarui, *Jpn. J. Appl. Phys.* **27**, 506 (1988).

126. K. Yamagishi and Y. Tarui, *Jpn. J. Appl. Phys.* **25**, L306 (1986).

127. B. J. Morris, *Appl. Phys. Lett.* **48**, 867 (1986).

128. P. D. Brewer, J. E. Jensen, G. L. Olson, L. W. Tutt, and J. J. Zinck, *Mater. Res. Soc. Symp. Proc.* **101**, 327 (1988).

129. S. J. C. Irvine, H. Hill, G. T. Brown, S. J. Barnett, J. E. Hails, O. D. Dosser, and J. B. Mullin, *J. Vac. Sci. Technol. B* **7**, 1191 (1989).

130. D. S. Alderdice, *J. Mol. Spectrosc.* **15**, 509 (1965).

131. C. A. L. Becker, C. J. Ballhausen, and I. Trabjerg, *Theor. Chim. Acta* **13**, 355 (1969).

132. P. Davidovits and J. A. Bellisio, *J. Chem. Phys.* **50**, 3560 (1969).

133. F. Pennella and W. J. Taylor, *J. Mol. Spectrosc.* **11**, 321 (1963).

134. F. A. Miller and W. B. White, *Spectrochim. Acta* **9**, 98 (1957).

135. S. J. C. Irvine, unpublished.

136. J. Perrin and T. Broekhuizen, *Appl. Phys. Lett.* **50**, 433 (1987).

137. Y. Matsui, A. Yuuki, N. Morita, and K. Tachibana, *Jpn. J. Appl. Phys.* **26**, 1575 (1987).

138. C.-H. Wu, *J. Phys. Chem.* **91**, 5054 (1987).

139. K. Kamisako, T. Imai, and Y. Tarui, *Jpn. J. Appl. Phys.* **27**, 1092 (1988).

140. C. J. Kiely, V. Tavitian, C. Jones, and J. G. Eden, *Appl. Phys. Lett.* **55**, 65 (1989).

141. L. Hellner, K. T. V. Grattan, and M. H. R. Hutchinson, *J. Chem. Phys.* **81**, 4389 (1984).

142. Y. Sato, K. Matsushita, T. Hariu, and Y. Shibata, *Appl. Phys. Lett.* **44**, 592 (1984).

143. S. Oda, R. Kawase, T. Sato, I. Shimizu, and H. Kokado, *Appl. Phys. Lett.* **48**, 33 (1986).

144. T. Y. Sheng, Z. Q. Yu, and G. J. Collins, *Appl. Phys. Lett.* **52**, 576 (1988).

145. I. Suemune, Y. Kunitsugu, Y. Tanaka, Y. Kan, and M. Yamanishi, *Appl. Phys. Lett.* **53**, 2173 (1988).

146. L. C. Lee, X. Wang, and M. Suto, *J. Chem. Phys.* **86**, 4353 (1987).

147. M. D. Person, K. Q. Lao, B. J. Eckholm, and L. J. Butler, *J. Chem. Phys.* **91**, 812 (1989).

148. K. J. Scoles, A. H. Kim, M.-H. Jiang, and B. C. Lee, *J. Vac. Sci. Technol. B* **6**, 470 (1988).

149. C. Fuchs, E. Boch, E. Fogarassy, B. Aka, and P. Siffert, *Mater. Res. Soc. Symp. Proc.* **101**, 361 (1988).

150. G. W. Tyndall, C. E. Larson, and R. L. Jackson, *J. Phys. Chem.* **93**, 5508 (1989).

151. P. A. Hackett and P. John, *J. Chem. Phys.* **79**, 3593 (1983).

152. P. A. Hackett and P. John, *J. Chem. Phys.* **79**, 4815 (1983).

153. S. A. Mitchell and P. A. Hackett, *Chem. Phys. Lett.* **107**, 508 (1984).

154. Y. Arai, S. Yamaguchi, and T. Ohsaki, *Appl. Phys. Lett.* **52**, 2083 (1988).

155. M. Sasaki, Y. Kawakyu, and M. Mashita, *Jpn. J. Appl. Phys.* **28**, L131 (1989).

156. P. S. Shaw, E. Sanchez, J. A. O'Neill, Z. Wu, and R. M. Osgood, *J. Chem. Phys.* **94**, 1643 (1991).

157. D. J. Ehrlich and R. M. Osgood, Jr., *Chem. Phys. Lett.* **79**, 381 (1981).

158. Y. Rytz-Froidevaux, R. P. Salathé, H. H. Gilgen, and H. P. Weber, *Appl. Phys. A* **27**, 133 (1982).

159. C. J. Chen and R. M. Osgood, Jr., *Chem. Phys. Lett.* **98**, 363 (1983); C. J. Chen and R. M. Osgood, Jr., *Mater. Res. Soc. Symp. Proc.* **17**, 169 (1983).

160. See, for example, H. Lüth, *J. Vac. Sci. Technol. A* **7**, 696 (1989).

161. D. J. Ehrlich, R. M. Osgood, Jr., and T. F. Deutsch, *Appl. Phys. Lett.* **38**, 946 (1981).

162. J. Y. Tsao and D. J. Ehrlich, *Appl. Phys. Lett.* **45**, 617 (1984).

163. G. S. Higashi and C. G. Fleming, *Appl. Phys. Lett.* **48**, 1051 (1986).

164. M. Shimizu, T. Katayama, Y. Tanaka, T. Shiosaki, and A. Kawabata, *J. Cryst. Growth* **101**, 171 (1990).

165. F. A. Houle, *J. Vac. Sci. Technol. B* **7**, 1149 (1989).

166. T. E. Orlowski and D. A. Mantell, *J. Vac. Sci. Technol. A* **7**, 2598 (1989).

167. G. S. Higashi and L. J. Rothberg, *Appl. Phys. Lett.* **47**, 1288 (1985).

168. M. Hanabusa, A. Oikawa, and P. Y. Cai, *J. Appl. Phys.* **66**, 3268 (1989).

169. D. Braichotte and H. van den Bergh, *Appl. Phys. A* **45**, 337 (1988).

170. Y. Zhang and M. Stuke, *J. Cryst. Growth* **93**, 143 (1988).

171. E. Villa, J. S. Horwitz, and D. S. Y. Hsu, *Chem. Phys. Lett.* **164**, 587 (1989).

172. K. Leggett, J. C. Polanyi, and P. A. Young, *J. Chem. Phys.* **93**, 3645 (1990).

173. W. L. Ahlgren, E. J. Smith, J. B. James, T. W. James, R. P. Ruth, and E. A. Patten, *J. Cryst. Growth* **86**, 198 (1988).

174. J. Watanabe and M. Hanabusa, *J. Mater. Res.* **4**, 882 (1989).

175. H. Zarnani, Z. Q. Yu, G. J. Collins, E. Bhattacharya, and J. I. Pankove, *Appl. Phys. Lett.* **53**, 1314 (1988).

176. Z. Yu and G. J. Collins, *Phys. Scr.* **41**, 25 (1990).

177. K. K. Schuegraf, in *Microelectronic Manufacturing and Testing*, pp. 1–4. Lake, Libertyville, 1983.

178. R. Solanki and G. J. Collins, *Appl. Phys. Lett.* **42**, 662 (1983).

179. J. J. Zinck, P. D. Brewer, J. E. Jensen, G. L. Olson, and L. W. Tutt, *Appl. Phys. Lett.* **52**, 1434 (1988).

180. A. A. Langford, J. Bender, M. L. Fleet, and B. L. Stafford, *J. Vac. Sci. Technol. B* **7**, 437 (1989).

181. "LACVD [Laser-Activated CVD] System Accommodates Different Sizes and Types of Substrates." *Semiconductor International*, pp. 198, 199 (April 1988).

182. Y. Nambu, Y. Morishige, and S. Kishida, *Appl. Phys. Lett.* **56**, 2581 (1990).

183. BHK, Inc., Monrovia, CA 91076.

184. B. Brehm and H. Siegert, *Z. Angew. Phys.* **19**, 244 (1965).

185. P. G. Wilkinson and E. T. Byram, *Appl. Opt.* **4**, 581 (1965).

186. H. Kumagai and M. Obara, *Appl. Phys. Lett.* **55**, 1583 (1989).

187. K. Yamada, K. Miyazaki, T. Hasama, and T. Sato, *Appl. Phys. Lett.* **54**, 597 (1989).

188. A. B. Petersen, *Proc. LEOS'89, 2nd Annu. Meet. IEEE Lasers Electro-Opt. Soc.*, Orlando, FL (October, 1989).

189. S. D. Baker, W. I. Milne, and P. A. Robertson, *Appl. Phys. A* **46**, 243 (1988).

190. T. Urisu, H. Kyuragi, Y. Utsumi, J. Takahashi, and M. Kitamura, *Rev. Sci. Instrum.* **60**, 2157 (1989).

191. T. F. Deutsch, D. J. Ehrlich, and R. M. Osgood, Jr., *Appl. Phys. Lett.* **35**, 175 (1979).

192. D. J. Ehrlich, R. M. Osgood, Jr., and T. F. Deutsch, *IEEE J. Quantum Electron.* **QE-16**, 1233 (1980).

193. R. W. Andreatta, C. C. Abele, J. F. Osmundsen, J. G. Eden, D. Lubben, and J. E. Greene, *Appl. Phys. Lett.* **40**, 183 (1982).

194. R. Solanki, W. H. Ritchie, and G. J. Collins, *Appl. Phys. Lett.* **43**, 454 (1983).

195. T. Motooka, S. Gorbatkin, D. Lubben, D. Eres, and J. E. Greene, *J. Vac. Sci. Technol. A* **4**, 3146 (1986).

196. D. A. Mantell, *Appl. Phys. Lett.* **53**, 1387 (1988).

197. M. Hanabusa, K. Hayakawa, A. Oikawa, and K. Maeda, *Jpn. J. Appl. Phys.* **27**, L1392 (1988).

198. T. H. Baum, E. E. Marinero, and C. R. Jones, *Appl. Phys. Lett.* **49**, 1213 (1986).

199. M. R. Aylett and J. Haigh, in *Extended Abstracts: Beam-Induced Chemical Processes* (R. J. von Gutfeld, J. E. Greene, and H. R. Schlossberg, eds.), Proc. Symp. D, Mater. Res. Soc. Fall 1985 Meeting, Boston.

200. M. W. Jones, L. J. Rigby, and D. Ryan, *Nature (London)* **212**, 177 (1966).

201. D. J. Ehrlich and R. M. Osgood, Jr., *Thin Solid Films* **90**, 287 (1982).

202. R. M. Osgood, Jr. and D. J. Ehrlich, *Opt. Lett.* **7**, 385 (1982).

203. S. R. J. Brueck and D. J. Ehrlich, *Phys. Rev. Lett.* **48**, 1678 (1982).

204. D. J. Ehrlich, R. M. Osgood, Jr., and T. F. Deutsch, *J. Vac. Sci. Technol.* **20**, 738 (1982).

205. T. H. Wood, J. C. White, and B. A. Thacker, *Appl. Phys. Lett.* **42**, 408 (1983).

206. R. Solanki, P. K. Boyer, J. E. Mahan, and G. J. Collins, *Appl. Phys. Lett.* **38**, 572 (1981).

207. R. Solanki, P. K. Boyer, and G. J. Collins, *Appl. Phys. Lett.* **41**, 1048 (1982).

208. H. Yokoyama, F. Uesugi, S. Kishida, and K. Washio, *Appl. Phys. A* **37**, 23 (1985).

209. T. M. Mayer, G. J. Fisanick, and T. S. Eichelberger, IV, *J. Appl. Phys.* **53**, 8462 (1982).

210. K. A. Singmaster, F. A. Houle, and R. J. Wilson, *Appl. Phys. Lett.* **53**, 1048 (1988).

211. N. S. Gluck, G. J. Wolga, C. E. Bartosch, W. Ho, and Z. Ying, *J. Appl. Phys.* **61**, 998 (1987).

212. C. R. Jones, F. A. Houle, C. A. Kovac, and T. H. Baum, *Appl. Phys. Lett.* **46**, 97 (1985).

213. F. A. Houle, R. J. Wilson, and T. H. Baum, *J. Vac. Sci. Technol. A* **4**, 2452 (1986).

214. P. M. George and J. L. Beauchamp, *Thin Solid Films* **67**, L25 (1980).

215. P. J. Love, R. T. Loda, P. R. LaRoe, A. K. Green, and V. Rehn, *Mater. Res. Soc. Symp. Proc.* **29**, 101 (1984).

216. N. Bottka, P. J. Walsh, and R. Z. Dalbey, *J. Appl. Phys.* **54**, 1104 (1983).

217. J. S. Foord and R. B. Jackman, *J. Opt. Soc. Am. B* **3**, 806 (1986).

218. J. V. Armstrong, A. A. Burk, Jr., J. M. D. Coey, and K. Moorjani, *Appl. Phys. Lett.* **50**, 1231 (1987).

219. M. R. Aylett and J. Haigh, *Mater. Res. Soc. Symp. Proc.* **17**, 177 (1983).

220. H. H. Gilgen, T. Cacouris, P. S. Shaw, R. R. Krchnavek, and R. M. Osgood, Jr., *Appl. Phys. B* **42**, 55 (1987).

221. L. J. Rigby, *Trans. Faraday Soc.* **65**, 2421 (1969).

222. D. L. Perry and M. W. Roberts, *J. Chem. Soc., Chem. Commun.* p. 147 (1972).

223. H. Schröder, I. Gianinoni, D. Masci, and K. L. Kompa, in *Chemical Vapor*

Deposition (K. F. Jensen, ed.). Academic Press, San Diego, CA (in press).

224. C. Garrido-Suarez, D. Braichotte, and H. van den Bergh, *Appl. Phys. A* **46**, 285 (1988); D. Braichotte, C. Garrido, and H. van den Bergh, *Appl. Surf. Sci.* **46**, 9 (1990).

225. W. E. Johnson and L. A. Schlie, *Appl. Phys. Lett.* **40**, 798 (1982).

226. S. P. Kowalczyk and D. L. Miller, *J. Appl. Phys.* **59**, 287 (1986).

227. J. Y. Tsao, R. A. Becker, D. J. Ehrlich, and F. J. Leonberger, *Appl. Phys. Lett.* **42**, 559 (1983).

228. R. L. Jackson and G. W. Tyndall, *J. Appl. Phys.* **62**, 315 (1987); **64**, 2092 (1988).

229. T. F. Deutsch and D. D. Rathman, *Appl. Phys. Lett.* **45**, 623 (1984).

230. A. E. Adams, M. L. Lloyd, S. L. Morgan, and N. G Davis, in *Laser Processing and Diagnostics* (D. Bäuerle, ed.), pp. 269–273. Springer-Verlag, Berlin, 1984.

231. T. F. Deutsch, D. J. Ehrlich, R. M. Osgood, Jr. and Z. L. Liau, *Appl. Phys. Lett.* **36**, 847 (1980).

232. R. R. Krchnavek, H. H. Gilgen, J. C. Chen, P. S. Shaw, T. J. Licafa, and R. M. Osgood, Jr., *J. Vac. Sci. Technol. B* **5**, 20 (1987).

233. M. R. Aylett and J. Haigh, in *Laser Processing and Diagnostics* (D. Bäuerle, ed.), p. 263. Springer-Verlag, Berlin, 1984.

234. H. Ando, H. Inuzuka, M. Konagai, and K. Takahashi, *J. Appl. Phys.* **58**, 802 (1985).

235. R. L. Jackson, *J. Chem. Phys.* **92**, 807 (1990).

236. R. Larciprete and E. Borsella, *Nuovo Cimento* **11D**, 1603 (1989).

237. G. S. Higashi, *Appl. Surf. Sci.* **43**, 6 (1989).

238. T. H. Baum, C. E. Larson, and R. L. Jackson, *Appl. Phys. Lett.* **55**, 1264 (1989).

239. C. Sasaoka, K. Mori, Y. Kato, and A. Usui, *Appl. Phys. Lett.* **55**, 741 (1989).

240. K. A. Singmaster, F. A. Houle, and R. J. Wilson, *J. Phys. Chem.* **94**, 6864 (1990).

241. L. Konstantinov, R. Nowak, and P. Hess, *Appl. Phys. A* **47**, 171 (1988).

242. R. Nowak, L. Konstantinov, and P. Hess, *Appl. Surf. Sci.* **36**, 177 (1989).

243. G. A. Kovall, J. C. Matthews, and R. Solanki, *J. Vac. Sci. Technol. A* **6**, 2353 (1988).

244. R. Nowak, P. Hess, H. Oetzmann, and C. Schmidt, *Appl. Surf. Sci.* **43**, 11 (1989).

245. F. A. Houle, *Laser Chem.* **9**, 107 (1988).

246. T. H. Baum, *J. Electrochem. Soc.* **137**, 252 (1990).

247. J. Haigh and M. R. Aylett, in *Laser Microfabrication: Thin Film Processes and Lithography* (D. J. Ehrlich and J. Y. Tsao, eds.), pp. 453–501. Academic Press, San Diego, CA, 1989.

248. T. Donohue, in *Laser Applications in Physical Chemistry* (D. K. Evans, ed.), pp. 89–172. Dekker, New York, 1989.

249. H. Tachibana, A. Nakaue, and Y. Kawate, *Mater. Res. Soc. Symp. Proc.* **101**, 367 (1988).

250. K. Kitahama, K. Hirata, H. Nakamatsu, S. Kawai, N. Fujimori, T. Imai, H. Yoshino, and A. Doi, *Appl. Phys. Lett.* **49**, 634 (1986).

251. K. Kitahama, K. Hirata, H. Nakamatsu, S. Kawai, N. Fujimori, and Y. Imai, *Mater. Res. Soc. Symp. Proc.* **75**, 309 (1987).
252. K. Kitahama, *Appl. Phys. Lett.* **53**, 1812 (1988).
253. K. Kitahama, K. Hirata, H. Nakamatsu, S. Kawai, N. Fujimori, T. Imai, H. Yoshino, and A. Doi, *Appl. Phys. Lett.* **54**, 968 (1989).
254. Y. Rousseau and G. J. Mains, *J. Phys. Chem.* **70**, 3158 (1966).
255. T. Saitoh, S. Muramatsu, S. Matsubara, and M. Migitaka, *Jpn. J. Appl. Phys.* **22**, Suppl. 22-1, 617 (1982); *Proc. 14th Conf. Solid State Devices* p. 617 (1982).
256. T. Saitoh, S. Muramatsu, T. Shimada, and M. Migitaka, *Appl. Phys. Lett.* **42**, 678 (1983).
257. T. Saitoh, T. Shimada, M. Migitaka, and Y. Tarui, *J. Non-Cryst. Solids* **59**, 715 (1983).
258. Y. Tarui, K. Sorimachi, K. Fujii, K. Aota, and T. Saito, *J. Non-Cryst. Solids* **59/60**, 711 (1983).
259. J. Dutta, A. L. Unaogu, S. Ray, and A. K. Barua, *J. Appl. Phys.* **66**, 4709 (1989).
260. K. Aota, Y. Tarui, and T. Saito, in *Amorphous Semiconductor Technologies and Devices* (Y. Hamakawa, ed.), pp. 98–107. Ohmsha, Tokyo and North-Holland, Amsterdam, 1984.
261. S. Mizukawa, K. Sato, K. Yasuhiro, M. Isawa, K. Kuroiwa, and Y. Tarui, *Jpn. J. Appl. Phys.* **28**, 961 (1989).
262. K. Suzuki, K. Kuroiwa, K. Kamisako, and Y. Tarui, *Appl. Phys. A* **50**, 227 (1990).
263. K. Kamisako, T. Imai, and Y. Tarui, *Jpn. J. Appl. Phys.* **27**, 1092 (1988).
264. T. Inoue, M. Konagai, and H. Takahashi, *Appl. Phys. Lett.* **43**, 774 (1983).
265. Y. Tarui, K. Aota, T. Sugiura, and T. Saitoh, *Mater. Res. Soc. Symp. Proc.* **29**, 109 (1984).
266. S. Nishida, H. Tasaki, M. Konagai, and K. Takahashi, *J. Appl. Phys.* **58**, 1427 (1985).
267. T. Tatsuya, W. Y. Kim, M. Konagai, and K. Takahashi, *Appl. Phys. Lett.* **45**, 865 (1984).
268. T. Tanaka, Y. K. Woo, M. Konagai, and K. Takahashi, *Appl. Phys. Lett.* **45**, 865 (1984).
269. R. E. Rocheleau, S. S. Hegedus, W. A. Buchanan, and S. C. Jackson, *Appl. Phys. Lett.* **51**, 133 (1987).
270. S. Nishida, T. Shiimoto, A. Yamada, S. Karasawa, M. Konagai, and K. Takahashi, *Appl. Phys. Lett.* **49**, 79 (1986).
271. Y. Jia, A. Yamada, M. Konagai, and K. Takahashi, *Jpn. J. Appl. Phys.* **30**, 893 (1991).
272. W. I. Milne, F. J. Clough, S. C. Deane, S. D. Baker, and P. A. Robertson, *Appl. Surf. Sci.* **43**, 277 (1989).
273. Y. Mishima, M. Hirose, Y. Osaka, K. Nagamine, Y. Ashida, N. Kitagawa, and K. Isogaya, *Jpn. J. Appl. Phys.* **22**, L46 (1983).
274. Y. Mishima, Y. Ashida, and M. Hirose, *J. Non-Cryst. Solids* **59/60**, 707 (1983).

275. Y. Mishima, M. Hirose, Y. Osaka, and Y. Ashida, *J. Appl. Phys.* **55**, 1234 (1984).

276. T. Wadayama, W. Suetaka, and A. Sekiguchi, *Jpn. J. Appl. Phys.* **27**, 501 (1988).

277. T. Fuyuki, K.-Y. Du, S. Okamoto, S. Yasuda, T. Kimoto, M. Yoshimoto, and H. Matsunami, *J. Appl. Phys.* **64**, 2380 (1988).

278. M. Kawasaki, Y. Tsukiyama, and H. Hada, *J. Appl. Phys.* **64**, 3254 (1988).

279. N. Gonohe, S. Shimizu, K. Tamagawa, T. Hayashi, and H. Yamakawa, *Jpn. J. Appl. Phys.* **26**, L1189 (1987).

280. T. Yamazaki, H. Minakata, and T. Ito, *J. Electrochem. Soc.* **137**, 1981 (1990).

281. N. Fujiki, Y. Nakatani, K. Inoue, M. Okuyama, and Y. Hamakawa, *Jpn. J. Appl. Phys.* **28**, 829 (1989).

282. K. Kumata, U. Itoh, Y. Toyoshima, N. Tanaka, H. Anzai, and A. Matsuda, *Appl. Phys. Lett.* **48**, 1380 (1986).

283. W.-Y. Kim, A. Shibata, Y. Kazama, M. Konagai, and K. Takahashi, *Jpn. J. Appl. Phys.* **28**, 311 (1989).

284. A. Yoshikawa and S. Yamaga, *Jpn. J. Appl. Phys.* **24**, 1585 (1985).

285. A. Yamada, M. Konagai, and K. Takahashi, *Jpn. J. Appl. Phys.* **24**, 1586 (1985).

286. H. Zarnani, H. Demiryont, and G. J. Collins, *J. Appl. Phys.* **60**, 2523 (1986).

287. Y. Hiura, H. Uchida, S. Kaneko, M. Morishige, and S. Kishida, NEC Corp., Tokyo (unpublished).

288. A. Yamada, A. Satoh, M. Konagai, and K. Takahashi, *J. Appl. Phys.* **65**, 4268 (1989).

289. S. Lian, B. Fowler, D. Bullock, and S. Banerjee, *Appl. Phys. Lett.* **58**, 514 (1991).

290. T. Tanaka, K. Deguchi, and M. Hirose, *Jpn. J. Appl. Phys.* **26**, 2057 (1987).

291. M. Ishida, H. Tanaka, K. Sawada, A. Namiki, T. Nakamura, and N. Ohtake, *J. Appl. Phys.* **64**, 2087 (1988).

292. R. G. Frieser, *J. Electrochem. Soc.* **115**, 401 (1968).

293. A. Ishitani, Y. Oshita, K. Tanigaki, and K. Takada, *J. Appl. Phys.* **61**, 2224 (1987).

294. J. G. Eden, J. E. Greene, J. F. Osmundsen, D. Lubben, C. C. Abele, S. Gorbatkin, and H. D. Desai, *Mater. Res. Soc. Symp. Proc.* **17**, 185 (1983).

295. V. Tavitian, C. J. Kiely, D. B. Geohegan, and J. G. Eden, *Appl. Phys. Lett.* **52**, 1710 (1988); V. Tavitian, C. J. Kiely, and J. G. Eden, *Mater. Res. Soc. Symp. Proc.* **101**, 349 (1988).

296. C. J. Kiely, V. Tavitian, and J. G. Eden, *J. Appl. Phys.* **65**, 3883 (1989).

297. H. H. Burke, I. P. Herman, V. Tavitian, and J. G. Eden, *Appl. Phys. Lett.* **55**, 253 (1989).

298. A. E. Stanley, R. A. Johnson, J. B. Turner, and A. H. Roberts, *Appl. Spectrosc.* **40**, 374 (1986).

299. S. J. C. Irvine, *Appl. Phys. A* **41**, 503 (1986).

300. T. Maruyama and A. Nakai, *Jpn. J. Appl. Phys.* **28**, L346 (1989).

301. S. Fujita, A. Tanabe, T. Sakamoto, M. Isemura, and S. Fujita, *Jpn. J. Appl. Phys.* **26**, L2000 (1987).

302. S. Fujita, F. Takeuchi, and S. Fujita, *Jpn. J. Appl. Phys.* **27**, L2019 (1988).

303. S. Fujita, A. Tanabe, T. Sakamoto, M. Isemura, and S. Fujita, *J. Cryst. Growth* **93**, 259 (1988).

304. M. Ohishi, H. Saito, H. Okano, and K. Ohmori, *J. Cryst. Growth* **95**, 538 (1989).

305. N. Matsumura, M. Tsubokura, K. Miyagawa, N. Nakamura, Y. Miyanagi, T. Fukada, and J. Saraie, *Jpn. J. Appl. Phys.* **29**, L723 (1990).

306. N. Matsumura, T. Fukada, K. Senga, Y. Fukushima, and J. Saraie, *J. Cryst. Growth* **111**, 787 (1991).

307. H. Ogawa, M. Nishio, M. Ikejiri, and M. Tuboi, *Appl. Phys. Lett.* **58**, 2384 (1991).

308. K. Hirabayashi and H. Kozawaguchi, *Jpn. J. Appl. Phys.* **28**, 814 (1989).

309. G. B. Shinn, P. M. Gillespie, W. L. Wilson, Jr., and W. M. Duncan, *Appl. Phys. Lett.* **54**, 2440 (1989).

310. S. J. C. Irvine, J. B. Mullin, and J. Tunnicliffe, *J. Cryst. Growth* **68**, 188 (1984).

311. S. J. C. Irvine, J. B. Mullin, and J. Tunnicliffe, *Springer Ser. Chem. Phys.* **39**, 234 (1984).

312. S. J. C. Irvine, J. B. Mullin, G. W. Blackmore, O. D. Dosser, and H. Hill, *Mater. Res. Soc. Symp. Proc.* **90**, 153 (1987).

313. S. J. C. Irvine, J. B. Mullin, H. Hill, G. T. Brown, and S. J. Barnett, *J. Cryst. Growth* **86**, 188 (1988).

314. W. I. Ahlgren, J. B. James, R. P. Ruth, E. A. Patten, and J. L. Staudenmann, *Mater. Res. Soc. Symp. Proc.* **90**, 405 (1987).

315. S. J. C. Irvine, H. Hill, J. E. Hails, J. B. Mullin, S. J. Barnett, G. W. Blackmore, and O. D. Dosser, *J. Vac. Sci. Technol. A* **8**, 1059 (1990).

316. D. W. Kisker and R. D. Feldman, *Mater. Lett.* **3**, 485 (1985).

317. D. W. Kisker and R. D. Feldman, *J. Cryst. Growth* **72**, 102 (1985).

318. N. W. Cody, U. Sudarsan, and R. Solanki, *J. Mater. Res.* **3**, 1144 (1988).

319. N. W. Cody, U. Sudarsan, and R. Solanki, *J. Appl. Phys.* **66**, 449 (1989).

320. B. Liu, R. F. Hicks, J. J. Zinck, J. E. Jensen, and G. L. Olson, *J. Vac. Sci. Technol. A* (to be published).

321. S. J. C. Irvine, J. B. Mullin, G. W. Blackmore, and O. D. Dosser, *J. Vac. Sci. Technol. B* **3**, 1450 (1985).

322. S. J. C. Irvine and J. B. Mullin, *J. Vac. Sci. Technol. A* **5**, 2100 (1987).

323. B. J. Morris, *Appl. Phys. Lett.* **48**, 867 (1986).

324. J. E. Jensen, P. D. Brewer, G. L. Olson, L. W. Tutt, and J. J. Zinck, *J. Vac. Sci. Technol. A* **6**, 2808 (1988).

325. N. Pütz, H. Heinecke, E. Veuhoff, G. Arens, M. Heyen, H. Lüth, and P. Balk, *J. Cryst. Growth* **68**, 194 (1984).

326. P. Balk, M. Fischer, D. Grundmann, R. Lückerath, H. Lüth, and W. Richter, *J. Vac. Sci. Technol. B* **5**, 1453 (1987); C. Plass, H. Heinecke, O. Kayser, H. Lüth, and P. Balk, *J. Cryst. Growth* **88**, 455 (1988).

327. J. Nishizawa, Y. Kokubun, H. Shimawaki, and M. Koike, *J. Electrochem. Soc.* **132**, 1939 (1985).

328. J. Nishizawa, H. Abe, and T. Kurobayashi, *J. Electrochem. Soc.* **132**, 1197 (1985).

329. J. Nishizawa, H. Shimawaki, and Y. Sakuma, *J. Electrochem. Soc.* **133**, 2567 (1986).

330. J. Nishizawa, T. Kurabayashi, H. Abe, and N. Sakurai, *J. Vac. Sci. Technol. A* **5**, 1572 (1987).

331. H. Kukimoto, Y. Ban, H. Komatsu, M. Takechi, and M. Ishizaki, *J. Cryst. Growth* **77**, 223 (1986).

332. S. S. Chu, T. L. Chu, C. L. Chang, and H. Firouzi, *Appl. Phys. Lett.* **52**, 1243 (1988).

333. P. K. York, J. G. Eden, J. J. Coleman, G. E. Fernández, and K. J. Beernink, *Appl. Phys. Lett.* **54**, 1866 (1989).

334. P. K. York, J. G. Eden, J. J. Coleman, G. E. Fernández, and K. J. Beernink, *J. Appl. Phys.* **66**, 5001 (1989).

335. Y. Kawakyu, H. Ishikawa, M. Sasaki, and M. Mashita, *Jpn. J. Appl. Phys.* **28**, L1439 (1989).

336. J. K. Ku, W. L. Choi, B. W. In , J. W. Chung, and O. D. Kwon, *GaAs on Si by Low Temperature Laser Chemical Vapor Deposition*, Paper UV2.2, Proc. 3rd Annu. Meet. IEEE Lasers Electro-Opt. Soc. (LEOS '90), Boston, MA, November 1990, p. 274.

337. T. Kachi, H. Ito, and S. Terada, *Jpn. J. Appl. Phys.* **27**, L1556 (1988).

338. D. P. Norton and P. K. Ajmera, *Appl. Phys. Lett.* **53**, 595 (1988).

339. D. P. Norton and P. K. Ajmera, *J. Electrochem. Soc.* **136**, 2371 (1989).

340. W. Roth, H. Kräutle, A. Krings, and H. Beneking, *Mater. Res. Soc. Symp. Proc.* **17**, 193 (1983).

341. Y. Aoyagi, S. Masuda, S. Namba, and A. Doi, *Appl. Phys. Lett.* **47**, 95 (1985).

342. A. Doi, Y. Aoyagi, and S. Namba, *Appl. Phys. Lett.* **48**, 1787 (1986); **49**, 785 (1986).

343. Y. Aoyagi, M. Kanazawa, A. Doi, S. Iwai, and S. Namba, *J. Appl. Phys.* **60**, 3131 (1986).

344. S. M. Bedair, J. K. Whisnant, N. H. Karam, M. A. Tischler, and T. Katsuyama, *Appl. Phys. Lett.* **48**, 174 (1986).

345. See, for example, N. H. Karam, H. Liu, I. Yoshida, B.-L. Jiang, and S. M. Bedair, *J. Cryst. Growth* **93**, 254 (1988), and references cited therein.

346. H. Sugiura, R. Iga, T. Yamada, and M. Yamaguchi, *Appl. Phys. Lett.* **54**, 335 (1989).

347. V. M. Donnelly, C. W. Tu, J. C. Beggy, V. R. McCrary, M. G. Lamont, T. D. Harris, F. A. Baiocchi, and R. C. Farrow, *Appl. Phys. Lett.* **52**, 1065 (1988).

348. C. W. Tu, V. M. Donnelly, J. C. Beggy, F. A. Baiocchi, V. R. McCrary, T. D. Harris, and M. G. Lamont, *Appl. Phys. Lett.* **52**, 966 (1988).

349. V. M. Donnelly and J. A. McCaulley, *Appl. Phys. Lett.* **54**, 2458 (1989).

350. R. Iga, H. Sugiura, T. Yamada, and K. Wada, *Appl. Phys. Lett.* **55**, 451 (1989).

351. K. Sugioka, K. Toyoda, K. Tachi, and M. Otsuka, *Appl. Phys. A* **49**, 723 (1989).

352. M. Mashita, *Jpn. J. Appl. Phys.* **28**, 690 (1989).

353. Y. Ban, M. Ishizaki, T. Asaka, Y. Koyama, and H. Kukimoto, *Jpn. J. Appl. Phys.* **28**, L1995 (1989).

354. J. J. Zinck, P. D. Brewer, J. E. Jensen, G. L. Olson, and L. W. Tutt, *Mater. Res. Soc. Symp. Proc.* **75**, 233 (1987).

355. U. Sudarsan, N. W. Cody, T. Dosluoglu, and R. Solanki, *Appl. Phys. Lett.* **55**, 738 (1989).

356. U. Sudarsan, N. W. Cody, T. Dosluoglu, and R. Solanki, *Appl. Phys. A* **50**, 325 (1990).

357. U. Sudarsan and R. Solanki, *J. Appl. Phys.* **67**, 2913 (1990).

358. P. C. John, J. A. Alwan, and J. G. Eden, *Thin Solid Films* (to be published).

359. V. M. Donnelly, M. Geva, J. Long, and R. F. Karlicek, *Appl. Phys. Lett.* **44**, 951 (1984).

360. V. M. Donnelly, D. Brasen, A. Appelbaum, and M. Geva, *J. Vac. Sci. Technol. A* **4**, 716 (1986).

361. M. Akazawa, H. Hasegawa, and E. Ohue, *Jpn. J. Appl. Phys.* **28**, L2095 (1989).

362. Y. K. Su, C. R. Huang, and Y. C. Chou, *Jpn. J. Appl. Phys.* **28**, 1664 (1989).

363. R. F. Sarkozy, in *Technical Digest of 1981 Symposium on VLSI Technology*, p. 68. Jpn. Soc. Appl. Phys., Hawaii, 1981.

364. W. H. Lan, S. L. Tu, S. J. Yang, and K. F. Huang, *Jpn. J. Appl. Phys.* **29**, 997 (1990).

365. R. Padmanabhan, B. J. Miller, and N. C. Saha, *Mater. Res. Soc. Symp. Proc.* **101**, 385 (1988).

366. K. P. Pande and V. K. R. Nair, *J. Appl. Phys.* **55**, 3109 (1984).

367. Y. Mishima, M. Hirose, Y. Osaka, and Y. Ashida, *J. Appl. Phys.* **55**, 1234 (1984).

368. Y. I. Nissim, J. L. Regolini, D. Bensahel, and C. Licoppe, *Electron. Lett.* **24**, 488 (1988).

369. M. Petitjean, N. Proust, and J.-F. Chapeaublanc, *Appl. Surf. Sci.* **46**, 189 (1990).

370. M. Niwano, S. Hirano, M. Suemitsu, K. Honma, and N. Miyamoto, *Jpn. J. Appl. Phys.* **28**, L1310 (1989).

371. M. Okuyama, Y. Toyoda, and Y. Hamakawa, *Jpn. J. Appl. Phys.* **23**, L97 (1984).

372. M. Okuyama, N. Fujiki, K. Inoue, and Y. Hamakawa, *Jpn. J. Appl. Phys.* **26**, L908 (1987).

373. K. Inoue, M. Okuyama, and Y. Hamakawa, *Jpn. J. Appl. Phys.* **27**, L2152 (1988).

374. M. Okuyama, N. Fujiki, K. Inoue, and Y. Hamakawa, *Appl. Surf. Sci.* **33/34**, 427 (1988).

375. H. Nonaka, K. Arai, Y. Fujino, and S. Ichimura, *J. Appl. Phys.* **64**, 4168 (1988).

376. K. Inoue, M. Nakamura, M. Okuyama, and Y. Hamakawa, *Appl. Phys. Lett.* **55**, 2402 (1989).

377. M. Nakamura, M. Okuyama, and Y. Hamakawa, *Jpn. J. Appl. Phys.* **29**, L687 (1990).

378. J. Marks and R. E. Robertson, *Appl. Phys. Lett.* **52**, 810 (1988).

379. P. K. Boyer, G. A. Roche, W. H. Ritchie, and G. J. Collins, *Appl. Phys. Lett.* **40**, 716 (1982).

380. P. K. Boyer, K. A. Emery, H. Zarnani, and G. J. Collins, *Appl. Phys. Lett.* **45**, 979 (1984).

381. T. Szorényi, P. González, M. D. Fernández, J. Pou, B. Léon, and M. Pérez-Amor, *Thin Solid Films* **193/194**, 619 (1990).

382. J. Watanabe and M. Hanabusa, *J. Mater. Res.* **4**, 882 (1989).

383. F. Houzay, J. M. Moison, and C. A. Sébenne, *Appl. Phys. Lett.* **58**, 1071 (1991).

384. H. M. Kim, S. S. Tai, S. L. Groves, and K. L. Schuegraf, in *Proceedings of the Eighth International Conference on Chemical Vapor Deposition*, Vol. 81–7, p. 258. Electrochemical Society, Pennington, NJ, 1981.

385. J. W. Peters, in *Technical Digest of the International Electron Devices Meeting 1981*, p. 240. Inst. Electr. Electron. Eng., New York, 1981.

386. T. Inushima, N. Hirose, K. Urata, T. Sato, and S. Yamazaki, *Appl. Surf. Sci.* **33/34**, 420 (1988).

387. P. K. Boyer, C. A. Moore, R. Solanki, W. K. Ritchie, G. A. Roche, and G. J. Collins, *Mater. Res. Soc. Symp. Proc.* **17**, 119 (1983).

388. T. Sugii, T. Ito, and H. Ishikawa, *Appl. Phys. A* **46**, 249 (1988).

389. E. Iborra, J. A. Lopez-Rubio, I. Esquivias, J. Sanz-Maudes, and T. Rodríguez, *J. Appl. Phys.* **67**, 1617 (1990).

390. T. F. Deutsch, D. J. Silversmith, and R. W. Mountain, *Mater. Res. Soc. Symp. Proc.* **29**, 67 (1984).

391. M. Ishida, A. Eto, T. Nakamura, and T. Suzaki, *J. Vac. Sci. Technol. A* **7**, 2931 (1989).

392. M. D. Hudson, C. Trundle, and C. J. Brierley, *J. Mater. Res.* **3**, 1151 (1988).

393. R. R. Kunz, M. Rothschild, and D. J. Ehrlich, *Appl. Phys. Lett.* **54**, 1631 (1989).

394. C. Arnone, M. Rothschild, J. G. Black, and D. J. Ehrlich, *Appl. Phys. Lett.* **48**, 1018 (1986).

395. M. Anpo, M. Sunamoto, and M. Che, *J. Phys. Chem.* **93**, 1187 (1989).

396. E. A. Mendoza, D. Sunil, E. Wolkow, H. D. Gafney, M. H. Rafailovich, J. Sokolov, G. G. Long, P. R. Jemian, S. A. Schwartz, and B. J. Wilkens, *Appl. Phys. Lett.* **57**, 209 (1990).

397. S. Motojima and H. Mizutani, *Appl. Phys. Lett.* **54**, 1104 (1989).

398. Y.-H. Jeong, G.-T. Kim, S.-T. Kim, K.-I. Kim, and W.-J. Chung, *J. Appl. Phys.* **69**, 6699 (1991).

399. H. Koinuma, K. A. Chaudhary, M. Nakabayashi, T. Shiraishi, T. Hashimoto, K. Kitazawa, Y. Suemune, and T. Yamamoto, *Jpn. J. Appl. Phys.* **30**, 656 (1991).

400. *Lambda Highlights*, No. 22, pp. 1–4. Lambda-Physik, Göttingen, 1990.

401. S. Motojima and H. Mizutani, *Appl. Phys. Lett.* **56**, 916 (1990).

402. J. Elders, D. Bebelaar and J. D. W. van Voorst, *Appl. Surf. Sci.* **46**, 215 (1990).

403. K. J. Mackey, D. C. Rodway, P. C. Smith and A. W. Vere, *Appl. Surf. Sci.* **43**, 1 (1990).

404. T. B. Stewart, G. S. Arnold, D. F. Hall, and H. D. Marten, *J. Phys. Chem.* **93**, 2393 (1989).

405. D. H. Maylotte and A. N. Wright, *Farady Discuss. Chem. Soc.* **58**, 292 (1975).

406. Y. Suzuki, *Jpn. J. Appl. Phys.* **28**, 920 (1989).

407. R. Putzar, H.-C. Petzold, and H. Staiger, *Appl. Surf. Sci.* **46**, 131 (1990).

408. A. Bauer, J. Ganz, K. Hesse, and E. Köhler, *Appl. Surf. Sci.* **46**, 113 (1990).

409. E. M. Young and W. A. Tiller, *Appl. Phys. Lett.* **50**, 80 (1987).

410. E. M. Young, *Appl. Phys. A* **47**, 259 (1988).

411. Z. Lu, M. T. Schmidt, D. V. Podlesnik, C. F. Yu, and R. M. Osgood, Jr., *J. Chem. Phys.* **93**, 7951 (1990), and references cited therein.

412. Y. Ishikawa, Y. Takagi, and I. Nakamichi, *Jpn. J. Appl. Phys.* **28**, L1453 (1989).

413. V. Crăciun, I. N. Mihăilescu, G. Oncioiu, A. Luches, M. Martino, V. Nassisi, E. Radiotis, A. V. Drigo, and S. Ganatsios, *J. Appl. Phys.* **68**, 2509 (1990).

414. W. G. Petro, I. Hino, S. Eglash, I. Lindau, S. Y. Su, and W. E. Spicer, *J. Vac. Sci. Technol.* **21**, 405 (1982).

415. I. V. Mitchell, G. Nyberg, and R. G. Elliman, *Appl. Phys. Lett.* **45**, 137 (1984).

416. C. J. Sofield, C. J. Woods, C. Wild, J. C. Riviere, and L. S. Welch, *Mater. Res. Soc. Symp. Proc.* **25**, 197 (1984).

417. A. J. Pedraza, M. J. Godbole, E. A. Kenik, D. H. Lowndes, and J. R. Thompson, Jr., *J. Vac. Sci. Technol. A* **6**, 1763 (1988).

418. C. Arnone, V. Daneu, and S. Riva-Sanseverino, *Appl. Phys. Lett.* **37**, 1012 (1980).

419. G. Auvert, D. Tonneau, and Y. Pauleau, *Appl. Phys. Lett.* **52**, 1062 (1988).

420. J. A. McClintock, R. A. Wilson, and N. E. Byer, *J. Vac. Sci. Technol.* **20**, 241 (1982).

421. J. L. Regolini, D. Bensahel, Y. I. Nissim, J. Mercier, E. Scheid, A. Perio, and E. Andre, *Electron. Lett.* **24**, 408 (1988).

422. Y. Takakuwa, M. Nogawa, M. Niwano, H. Katakura, S. Matsuyoshi, H. Ishida, H. Kato, and N. Miyamoto, *Jpn. J. Appl. Phys.* **28**, L1274 (1989).

423. S. J. Pearton, F. Ren, C. R. Abernathy, W. S. Hobson, and H. S. Luftman, *Appl. Phys. Lett.* **58**, 1416 (1991).

424. Y. Gosho, M. Yamada, and M. Saeki, *Jpn. J. Appl. Phys.* **29**, 950 (1990).

425. J. V. Mantese, A. B. Catalan, A. H. Hamdi, and A. L. Micheli, *Appl. Phys. Lett.* **52**, 1741 (1988).

426. J. V. Mantese, A. B. Catalan, A. M. Mance, A. H. Hamdi, A. L. Micheli, J. A. Sell, and M. S. Meyer, *Appl. Phys. Lett.* **53**, 1335 (1988).

427. S. Fujita, A. Tanabe, T. Kinoshita, and S. Fujita, *J. Cryst. Growth* **101**, 48 (1990).

428. M. Kitagawa, Y. Tomomura, K. Nakanishi, A. Suzuki, and S. Nakajima, *J. Cryst. Growth* **101**, 52 (1990).

429. M. Tsuji, N. Itoh, and Y. Nishimura, *Jpn. J. Appl. Phys.* **30**, 2868 (1991).

INDEX

191

M